Imagining Wild America

Imagining
Wild America

John R. Knott

Ann Arbor

The University of Michigan Press

Copyright © by the University of Michigan 2002
All rights reserved
Published in the United States of America by
The University of Michigan Press
Manufactured in the United States of America
⊗ Printed on acid-free paper

2005   2004   2003   2002   4   3   2   1

A CIP catalog record for this book is available from the British Library.

Library of Congress Cataloging-in-Publication Data

Knott, John Ray, 1937–
     Imagining wild America / John R. Knott.
          p.   cm.
     Includes index.
     ISBN 0-472-09806-3 (cloth : alk. paper) — ISBN 0-472-06806-7
(paper : alk. paper)
     1. American literature—History and criticism. 2. Nature in
literature. 3. Wilderness areas in literature. I. Title.
PS163 .K58   2002
810.9'36—dc21   2001007077

For the coming generation:
Josh, Dan, Jessica, Hannah, Eric,
Rose, Emma, Sophie

Contents

Acknowledgments

I am grateful to the Office of the Vice President for Research of the University of Michigan for a Michigan Humanities Award that gave me a semester of research leave and to the College of Literature, Science, and the Arts for grants that supported travels to libraries to examine manuscripts. Daryl Morrison of the University of the Pacific Library and Ron Limbaugh, then Director of the John Muir Center for Regional Studies, kindly assisted my work with the John Muir Papers. I am grateful also to the staffs of the Beinecke Library of Yale University and of the libraries of the American Museum of Natural History and the New York Historical Society for sharing their manuscripts of portions of Audubon's *Ornithological Biography* and to the Smithsonian Institution Libraries for providing photocopies of other Audubon manuscripts.

My friend and former chair, Bob Weisbuch, encouraged me to pursue my inclination to write this book, and my friend David Robertson gave welcome advice on developing the course on literature of the American wilderness from which it grew; he also extended an invitation to speak to the Nature and Culture Program of the University of California, Davis, which provided the stimulus for the initial version of the chapter on Edward Abbey. Conferences of the Association for the Study of Literature and the Environment (ASLE) at Missoula, Montana, and Kalamazoo, Michigan, provided responsive audiences for presentations on John James Audubon and Mary Oliver. Early conversations with Randall Roorda, then my student and course assistant, helped to shape the project, as did the comments of my colleagues James McIntosh, Linda Gregerson, and Keith Taylor on portions of the manuscript. John Tallmadge and anonymous readers for the University of Michigan Press provided important suggestions for revision. My friend Joe Vining contributed probing comments on the introduction. LeAnn Fields, my editor at the press, was strongly supportive and provided consistently good advice. I am also grateful to Carol Sickman-Garner for an exceptional job of copyediting.

An earlier version of chapter 4 appeared in *Western American Literature* 30 (1996); chapter 5 is an expanded version of an essay that appeared in *Essays in Literature* 23 (1996). I am grateful for permission to reprint this material. Excerpts from Muir's manuscripts are reproduced by permission of the John Muir Papers, Holt-Atherton Department of Special Collections. University of the Pacific Libraries. Copyright 1984 Muir-Hanna Trust.

As ever, I am deeply grateful to my wife, Anne Percy Knott, for her continuing interest and support.

Preface

This book grew out of the experience of teaching a course on the literature of the American wilderness for over a decade, during a period when many courses in what has come to be known as literature and the environment were sprouting on American campuses. I came to such teaching, like many others, through a combination of recreational reading and recreation. In my case, the recreation took the form of canoe-tripping with my wife, Anne, mainly in the Temagami area of Ontario, some five hours north of Toronto. In the course of reading what I thought of as "nature writing" during such trips and subsequent stays in a cabin on a small island in Lake Temagami, I asked myself why many of the works I was reading were not taught, in my university at least, and concluded that I should try teaching them myself. At the time, I was completing a book on Protestant martyrdom, a continuation of previous research on nonconformist literature in sixteenth- and seventeenth-century England, and feeling that I needed a respite from years of working with Puritan texts. So I launched a course on wilderness literature, hoping that there would be a constituency for it, and have been teaching it ever since.

The reasons that led me to offer such a course and, in time, other courses on literature and the environment are no doubt complex, but they have a lot to do with my own experiences of wild places. Two such experiences stand out in my memory. A particularly formative one was an eighteen-day canoe trip in the mid-1980s down the Noatak River in northern Alaska, beginning in the foothills of the Brooks Range and ending at the village of Noatak, near where the river empties into the Chukchi Sea. Most of the country we passed through would qualify as wilderness by any definition. We saw relatively little indication of human presence and abundant evidence of the animals whose territories we had entered, from the wolf den near our first campsite to the many tracks covering the gravelly beaches where we camped on our way down the river. We might have a fox come into our camp or caribou swim the river in front of our canoes. Caught up by the swift current, we swept past valleys that looked as though they could have been unexplored. From our campsites, once we were out of the mountains, we had long unobstructed vistas of tundra. When we ventured away from camp over the tussocky ground, perhaps in search of blueberries, we were wary of encountering a grizzly, the dominant presence in that landscape. My exposure to the backcountry of Alaska was brief, and superficial compared to that of many, but it gave me a powerful image of wilderness and a conviction about the value of such a place, even for those who will never see it.

My memories of the Temagami country of Ontario are more numerous, extending over more than twenty-five years, and I have developed a deep sense of connectedness with its Canadian shield landscape of granite shorelines, boreal forest, and clear cold water. One memory strikes me as particularly revealing. On one of our early trips in the area Anne and I were dropped by float plane with our canoe and supplies into the middle of beautiful, remote Florence Lake to begin a trip down the Lady Evelyn River. We paddled to an inviting sand beach to orient ourselves and begin to absorb our seemingly pristine setting. Before very long, as we were sinking into a feeling of solitude, we were surprised by a party of four canoes whose leader informed us that they had just taken the best campsite on the lake, on a point two hundred yards from where we were. It was one of many lessons in the fact that an area that epitomized the North Woods to us was in truth much used: by canoe campers; by loggers who cut up to the shoreline reserve in what we learned was crown forest; by the Bear Island Band of Anishnabai, who continue to hunt and trap in the area and to pursue such modern enterprises as building and maintaining places for summer visitors on the islands of Lake Temagami. We saved this particular occasion by paddling to the other end of the lake and finding a splendid high island where we spent the evening enjoying the profound silence of the lake, broken only by the calls of loons and other night sounds. It did not prove difficult to find what we were looking for, even if others had found it before and shared it with us.

Both experiences have influenced my teaching and the writing of this book. The first belongs to an American tradition of valuing and wanting to experience what we have come to understand as wilderness. Much of the watershed of the Noatak River, like various other areas of Alaska, can serve as an image of the possibility of wilderness, understood as a place at best lightly inhabited by humans. Perhaps the best example is the coastal plain of the Arctic National Wildlife Refuge, the endpoint for my course on wilderness literature since I began offering it. The fierce political battle over whether the coastal plain should be opened up to drilling for oil, which has continued at varying levels of intensity since the refuge was established in 1980 and has made this our most visible and symbolic example of wilderness, illustrates another American tradition—valuing natural areas primarily for the resources that can be extracted from them for human use. At a time when the idea of wilderness is under political attack from one direction and under intellectual attack from another, it is important to make a case for its continuing value, and one of the best ways to do this is to study writers who have enlarged its meaning and explored its implications in seeking to understand and represent their own experience.

If a place such as the coastal plain of the Arctic National Wildlife Refuge illustrates an ideal—an intact ecosystem with rich biological diversity and light human use (chiefly by the indigenous Gwich'n people, who pursue their traditional occupation of hunting the migrating caribou)—the Temagami area can stand for a more familiar reality, a landscape where nature and culture are complexly interrelated and the human presence is insistent. Most are likelier to experience the latter than the former, finding whatever wildness they encounter in an urban environment or in semi-rural settings close to home. It is also important to recognize that wildness does not have to be associated with places that could be defined as wilderness, that it persists in us and in the largely domesticated spheres in which we live, that in fact wildness and domesticity are interwoven. Many of the same writers who have had the most revealing things to say about the idea and the experience of wilderness are our best guides to the idea of wildness, a more supple and capacious term and one that continues to become more suggestive. This book and the course from which it evolved reflect my sense that both are worth attention.

Abbreviations

AP   Mary Oliver, *American Primitive* (Boston: Little Brown, 1983)

AR   Edward Abbey, *Abbey's Road* (New York: E. P. Dutton, 1979)

BP   Mary Oliver, *Blue Pastures* (New York: Harcourt Brace, 1995)

BW   Edward Abbey, *Beyond the Wall* (New York: Holt Rinehart, 1984)

Conf.   Edward Abbey, *Confessions of a Barbarian* (Boston: Little Brown, 1994)

CP   Wendell Berry, *Collected Poems, 1957–1982* (San Francisco: North Point Press, 1984)

DR   Edward Abbey, *Down the River* (New York: E. P. Dutton, 1982)

DS   Edward Abbey, *Desert Solitaire* (New York: Ballantine, 1968)

HE   Wendell Berry, *Home Economics* (San Francisco: North Point Press, 1987)

HL   Mary Oliver, *House of Light* (Boston: Beacon Press, 1990)

JH   Edward Abbey, *The Journey Home* (New York: E. P. Dutton, 1977)

LC   Mary Oliver, *The Leaf and the Cloud* (Cambridge, Mass.: Da Capo Press, 2000)

LLH   Wendell Berry, *The Long-Legged House* (New York: Harcourt, Brace, 1969)

MFS   John Muir, *My First Summer in the Sierra* (New York: Viking, 1987)

MW   Henry David Thoreau, *The Maine Woods* (Princeton: Princeton University Press, 1972)

NSP   Mary Oliver, *New and Selected Poems* (Boston: Beacon Press, 1992)

OB	John James Audubon, *Ornithological Biography,* 5 vols. (Edinburgh: Adam and Charles Black, 1831–39)
OL	Edward Abbey, *One Life at a Time* (New York: Henry Holt, 1988)
UW	Wendell Berry, *The Unforeseen Wilderness,* rev. ed. (San Francisco: North Point Press, 1991)
WP	Mary Oliver, *White Pine* (New York: Harcourt Brace, 1994)
WW	Mary Oliver, *West Wind* (Boston: Houghton Mifflin, 1997)

Introduction

John Muir's exuberant descriptions of the "fresh unblighted, unredeemed wilderness" that he found in his explorations of the Sierra
Nevada mountains of California popularized an ideal that has
shaped American thinking about the value of wilderness and the
importance of preserving it. They reflect a revolution in sensibility
influenced by English romantic writers and American transcenden-
talists, most notably Henry David Thoreau, by which wilderness
came to be seen as desirable, even as a manifestation of the sublime.
William Bradford's famous characterization of the Cape Cod found
by the settlers who arrived on the Mayflower as "a hideous and desolate wilderness, full of wild beasts and wild men," reflects a much
older sense of wilderness, going back to the desert wildernesses of
the Old Testament, as an inhospitable and dangerous place.[1] In his
story "Young Goodman Brown" Nathaniel Hawthorne captured the
Puritan sensibility in which the dark forest, the wilderness of the
early settlers, became a frightening and disorienting place of evil,
haunted by demonic Indians and the devil himself. By the time Muir
wrote, in the later nineteenth century, the appeal of wilderness as a
distinctive feature of the American landscape was firmly established. Muir could see the Sierra as a "range of light" and a vibrant,
pure, "divine wilderness" ordered and given life by a benevolent
God. If Muir's particular religion of nature is no longer so likely to
be shared, he nonetheless remains a cultural icon, widely quoted

and celebrated as the prophet of wilderness preservation and the first president of the Sierra Club. His writing, along with that of such other famous defenders of wilderness as Thoreau and Aldo Leopold and Edward Abbey, can be found in the *Trailside Reader* of the Sierra Club, a pocket-sized book of inspirational reading for backpackers.[2] Reading Muir and others who have meditated on the meaning of wild places has become a part of the American experience of wildness.

For all the popular fascination with wilderness, which increased dramatically in the later twentieth century, "wilderness" has in recent years become a contested and hence problematic term. Wilderness has long seemed an alien concept to Native Americans, a European import that served white culture as a way of signaling the strangeness of a natural world that indigenous peoples found familiar and sustaining, in fact regarded as home. More recently, Third World critics have attacked the notion of wilderness as an embodiment of a peculiarly American set of attitudes symbolized by a national park ideal that they see as inappropriate for countries in which intense human pressures on available land make preservation seem a luxury.[3] In India and Brazil, for example, critics have advocated "social ecology," a theory of conservation based upon preserving the living patterns of indigenous peoples, in opposition to the emphasis of conservation biologists upon preserving biological diversity.

Another important critique of the idea of wilderness, more relevant to my concerns in this book, has come from environmental historians and others who profess support for preserving wild areas but object to what they see as a pervasive habit of opposing nature and culture and consequently neglecting the role of humans in shaping and continuing to live with the natural world. I am thinking particularly of William Cronon's influential "The Trouble with Wilderness; or, Getting Back to the Wrong Nature" and other essays in the collection he edited, *Uncommon Ground: Toward Reinventing Nature* (1995). Michael Pollan's *Second Nature* (1991) contributed to the reconsideration of the contemporary American attraction to wilderness, which he sees as supported by a "wilderness ethic" deriving ultimately from Thoreau and Muir and "a romantic, pantheistic idea of nature that we invented in the first place."[4] Recent books by Susan G. Davis on the version of "nature" presented by SeaWorld and by Jennifer Price on such phenomena as the vogue of the plastic pink flamingo and the greening of television offer revealing commentaries on the ways in which we invent versions of nature that serve our various purposes.[5]

Those who have emphasized the constructedness of American views of wilderness and of nature more generally have focused on the consequences of what Price characterizes as patrolling the boundary between "Nature and non-Nature" and regarding nature as something "Out There."[6] They object to a tendency to remove humans from the natural environment and idealize a pristine nature, observing that the landscapes that European settlers found were shaped by millennia of human occupation.[7] Such critiques have led to a more complex understanding of how culturally inflected notions of wilderness and of nature have shaped attitudes toward the natural world. They offer a healthy corrective to some of the excesses of what Cronon calls "the ideology of wilderness" and make it difficult to ignore the historical and cultural contexts of American attitudes toward wilderness. Yet arguments against an "ideology of wilderness" or a "wilderness ethic" have the effect of discrediting any use of the term "wilderness," as do related arguments against subscribing to the "myth" of wilderness (or a pristine nature), often equated with the "myth" of Eden. Arguments of this sort frequently exaggerate the degree of human alteration of American landscapes prior to European settlement, which was significant in some places but in others minimal or nonexistent.[8]

Such arguments also tend to posit a monolithic community of advocates of wilderness preservation and to attribute to this community the ideal of a pristine, static nature removed from human influence. In fact, one finds a spectrum of beliefs among environmentalists about how possible and desirable it is to think about wilderness apart from past and future human presence. Many accept contemporary biological thinking about the dynamism of ecosystems and recognize the interdependency of nature and culture. The backlash against arguments of the sort made by Cronon has been motivated partly by political concerns of those engaged in ongoing battles for wilderness preservation and other forms of environmental protection. Biologist Michael Soulé has complained that the "social siege of nature" undermines efforts to resist a physical assault being carried out by "bulldozers, chainsaws, plows, and livestock."[9] A more fundamental criticism is that the preoccupation with cultural and historical perspectives ignores the biological reality of what Soulé would see as a variable "living nature" that it is inappropriate to think of as ever having been "virgin," since it was never static. Gary Snyder recognizes that wilderness is "in one sense a cultural construct" but faults critics for lacking "the awareness that wilderness is the locus of big rich ecosystems and is thus (among other things) a living place for beings who can survive in no other sort of habitat."[10]

My concern here is with the implications of the attack on the idea of wilderness and of a preoccupation with the projection of cultural attitudes in our understanding of the natural world for a body of literature in which wild nature is seen as a source of value. If the imaginative response to wilderness is simply a reflection of the influence of the romantic sublime and the myth of the vanishing frontier and a "flight from history," as Cronon seems to suggest, how does one make a case for reading Thoreau—or Edward Abbey, to take a more contemporary example? If value is located primarily in an awareness of history and of the demands of living in a highly developed society, how are we to regard a literature that finds meaning and personal restoration through close observation of the natural world?

The recent flourishing of critical writing about literature and the environment, commonly called ecocriticism, goes a long way toward answering questions about why we should read Thoreau, Abbey, and a great many others identified as belonging to a tradition of American nature writing. Lawrence Buell's *The Environmental Imagination* (1995) has done more than any other single work to date to define and enlarge this tradition, provide theoretical bases for ecocriticism, and stake out critical categories.[11] The influential *Ecocriticism Reader* (1996) and subsequent collections of critical essays have stimulated a lively debate about the meaning and viability of ecocriticism, while tracing its development as a critical approach.[12] Along with an increasing number of monographs, they have extended the range of the field, bringing ecocritical perspectives to canonical texts and reviving neglected ones, in the process giving fresh attention to regional literature and the burgeoning genre of nonfiction writing about place. Some of the most recent ecocritical writing challenges the habit of thinking of nature as something separate from and opposed to culture. Kent Ryden, for example, breaks down the boundaries we tend to draw between these categories by showing how landscapes that he grew up thinking of as wild nature, along with other New England landscapes, were in fact shaped by past land-use practices and cultural attitudes. John Tallmadge challenges these boundaries in a different way by questioning the implicit "Edenic" ideal that he sees as guiding the practice of ecological restoration; he argues for appreciating and learning from the "flourishing hybrid community" of alien and native species that he finds in the urban landscapes of Cincinnati.[13] An increasing interest in "urban nature" is one manifestation of ecocriticism's growing inclusiveness.

My purpose in this book is to turn attention back to the efforts of some of our best writers to imagine wild America, stimulated by various kinds of encounters with the natural world. One of my aims is to rehabilitate a vigorous tradition of writing about wilderness and wildness—an important aspect of the larger tradition of American nature writing—and to argue the value of these terms. They continue to have resonance for many, from frontline activists for wilderness preservation to those seeking to experience something they think of as "wild" nature beyond the perimeters of the thoroughly domesticated landscapes in which most Americans live, whether in remote places or in patches of wild nature close at hand. And they continue to be useful lenses through which to examine the allure of the natural world for writers in this tradition and their efforts to represent their relationship with this world.

I have chosen to exemplify the tradition by the work of six writers representing different stages and dimensions of the American fascination with wilderness and wildness: John James Audubon, Henry David Thoreau, John Muir, Edward Abbey, Wendell Berry, and Mary Oliver. Some of these choices, Thoreau and Muir in particular, seem inevitable for a project of this nature. Others, such as Abbey and Berry, reflect personal judgments about importance and influence. I represent the beginnings of the tradition with Audubon rather than with William Bartram and his famous *Travels* (1791) because Audubon's massive *Ornithological Biography* (1839) is still relatively unfamiliar and offers revealing glimpses of the frequently contradictory attitudes toward wilderness that one finds in early nineteenth-century America. The work of Oliver, the only poet in the group, shows the influence of the tradition I have described, particularly as this is embodied by Thoreau, but extends and redefines it in provocative and illuminating ways. Others might have been included: Mary Austin, Aldo Leopold, Gary Snyder, A. R. Ammons, Barry Lopez, Annie Dillard, Terry Tempest Williams, to name a few. Those that I have chosen seem to me both exemplary and influential. All have considerable popular reputations. Collectively, they represent the evolution of ways of imagining wilderness and wildness in America.

The seminal influence of Thoreau's writing about his own experience of wild nature and the implications of wildness makes him the central figure in my gallery of writers. Muir devoured Thoreau's writing, absorbed his influence, and embraced wildness with an enthusiasm and a physicality that go beyond anything one can find in Thoreau himself. Abbey conducted a good-natured quarrel with him in "Down the River with Henry Thoreau" and found his own ways of adapting Thoreau's critique of conventional economic and cultural attitudes, in his anarchic individualism enacting a version of Thoreau's defense of "absolute freedom and wildness." Berry has declared his allegiance to the "way of Thoreau," which he understands as involving an attraction to nature as a source of enlightenment and restoration. His moralizing commentary on society, grounded in convictions about the importance of regarding nature as valuable for its own sake and as embodying values that can guide human action, has prompted comparisons with Thoreau.[14] Oliver echoes Thoreau and imitates him in her habit of meditating on familiar places in a known landscape that she continually revisits in her walks. Thoreau's reflections on wilderness and wildness constitute only one strain in a complex body of writing, but this aspect of his work has been unusually consequential.

I have coupled wilderness and wildness when in fact "wild-ness" is a richer and more inclusive term. Wildness can be found in suburbia as well as in wilderness areas and can be seen as a property of body or mind.[15] Jack Turner has observed that "since wilderness is a place, and wildness a quality," we can ask how wild a "wilder-ness" or the experience of that wilderness is.[16] It makes sense to talk about degrees of wildness.[17] The term "wildness" has not attracted the kind of criticism that "wilderness" has, largely because the notion of wildness is more flexible and allows for the interaction of humans with the natural world.[18] Criticism of the term "wilder-ness" often focuses on definitions of the sort included in the federal Wilderness Act of 1964, with their implication that it is possible to preserve natural areas in something close to a pristine state by excluding any lasting human presence.

> A wilderness, in contrast with those areas where man and his own works dominate the landscape, is hereby recognized as an area where the earth and its commu-nity of life are untrammeled by man, where man himself is a visitor who does not remain. An area of wilderness is further defined to mean in this Act an area of undevel-oped Federal land retaining its primeval character and influence, without permanent improvements or human habitation.[19]

Wildness can be found in areas that could not be said to be "untrammeled" (unfettered) or "primeval," if any place can be said to be free of human influence when humans have altered the earth's climate, compromised the quality of its air, and precipitated the widespread movement of plant and animal species into bioregions to which they are not native. When Thoreau made his famous dec-laration, "In Wildness is the preservation of the world," he was thinking among other things of the wildness he discovered in his forays in the settled country around Concord, as well as in the wilderness of the Maine woods.[20] Such wildness, nowhere more apparent to him than in the "impervious" swamps he sought out in the familiar landscapes of his daily walks, possessed a vitality that he found both nourishing and a stimulus to unconventional thought and behavior. Wildness was "tonic" for Thoreau and, in varying ways, for the other writers whose works I discuss. All of these writers reveal an attraction to the idea of wilderness, some more than others, but for all of them but Thoreau's predecessor Audubon wildness is a more fundamental and pervasive ideal. For them the experience of wild nature, intimately and acutely observed, can be enlightening and liberating.[21]

I will argue that the tradition of writing about wilderness and wildness in America is in some respects, and especially for some writers, a visionary tradition that embraces values consciously understood to be ahistorical, values that cannot be accounted for simply by appeals to cultural evolution. Such writing frequently aspires to a sense of timelessness that depends on disengagement from a world of social habits and constraints and immersion in a natural order seen as prior to and more enduring than the human one. Writers in this tradition may see this natural order as having a mythic dimension and oppose its truth to that of history, as Thoreau does in his reflections on the quickly fading colors of the trout that he and his companions catch in his first venture into the Maine woods: "I could understand better, for this, the truth of mythology, the fables of Proteus, and all those beautiful sea-monsters—how all history, indeed put to a terrestrial use, is mere history; but put to a celestial, is mythology always."[22] Thoreau instinctively looked to classical mythology in his effort to understand a transcendent, "celestial" beauty that he saw as emblematic of primitive nature. Gary Snyder opposes a more generalized sense of myth to history in describing the initial European experience of the American West: "There is an almost invisible line that a person of the invading culture could walk across: out of history and into a perpetual present, a way attuned to the slower and steadier processes of nature. The possibility of passage into that myth-time world had been all but forgotten in Europe."[23] What appeals to Snyder is the possibility of reentering this "myth-time" world, whereas Thoreau is struck by how quickly the trout's beauty fades when it is out of the water and on its way to the frying pan—when it enters human history, so to speak. Both appeal to myth, however, to distinguish a natural order that they see as existing outside of what we think of as human history.

Whether or not they see themselves as pursuing a truth that they associate with myth, the writers with whom I am concerned find satisfaction in the intimate knowledge of particular natural environments and the cycles that govern life in them. They value a sense of living in a present that they associate with the natural world, a condition of being that depends upon shutting out the pre-occupations of everyday life and developing a heightened alertness to natural phenomena. Achieving this condition can involve an effort to quiet the "noise" of language and experience the "silence" of nonhuman nature, as it does in different ways for Berry and Oliver. This effort assumes the importance and the revelatory power—and the fundamental mysteriousness—of the natural order and also a resilience that enables this order to resist and even-tually outlast the various forms of human order that we may impose upon it.

When I describe writing as aspiring to a sense of timelessness, I mean that it aspires to an awareness of freedom from the time by which we measure daily life, from the ticking of the clock that reminds us of obligations implicit in the society to which we belong. And, frequently, from the timescale of recent human history. Other conceptions of time can suggest different perspectives and other ways of thinking about how we experience time: the dreamtime of Australian aborigines, the river time of Abbey's trip down the Col-orado, the cyclical time measured by the seasonal changes that so deeply engaged Thoreau, the geological time that Muir observed in the shaping of Sierra landscapes by glaciers. Aldo Leopold effec-tively juxtaposes geological and human timescales in *A Sand County Almanac* (1949), for example, in his rendering of the evolution of a Wisconsin marsh visited by sandhill cranes for millennia in "Marsh-land Elegy." In the perspective that he establishes by showing the slow formation of the peat that creates a bog and by making the crane, with its origins in the Eocene, "the symbol of our untamable past," the incursion of Europeans appears a very recent phenome-non. With deliberate casualness about dating events he describes a French trapper as appearing "one year not long ago" and the English as coming in their wagons "a century or two later."[24] Recent history proves disproportionately damaging in Leopold's story of the marsh, as we see settlers that follow the English attempting to drain it for cultivation and unintentionally igniting smoldering peat fires that they cannot extinguish. The juxtaposition of this history with the ancient past of the marsh and the more ancient lineage of the cranes reinforces his fundamental argument that "the ultimate value of these marshes is wildness, and the crane is wildness incarnate."[25] And the corollary that our sense of this wildness has nothing to do with familiar human timescales.

Seeking to understand and learn from natural orders does not have to mean ignoring human ones or the ways in which they condition our perceptions. Some of the writers I consider (Abbey, Berry, Oliver) are more sensitive than others to the limitations of language and to how it colors our seeing. All recognize ways in which landscapes are continually being transformed by human activity, even as they seek a dynamic in the natural world that is largely independent of this activity. Audubon, like other early naturalists, registered the losses he saw as settlement radically altered landscapes he had known in a wilder state, accepting these as the inevitable cost of progress while recording the unchanging rituals of bird behavior. Thoreau developed a sophisticated awareness of the ways in which European settlement had transformed and was continuing to transform landscapes around Concord, while looking for the "primitive nature" he valued most. Muir became a national leader in the fight for wilderness preservation, driven by concern over the increasing threats to the Sierra wilderness he knew intimately from years of exploration and study, yet he could minimize past human impacts in describing what he wanted to see as pure, dynamic wilderness manifesting a divine creative force. Assertions of the purity of wilderness were necessary to the rhetorical strategy he adopted in defending it. Abbey and Berry engage the historical forces they see as threatening their visions of the natural world, assuming the stances of satirist and prophet to attack abuses that they attribute to contemporary economic attitudes—primarily from development and "industrial tourism" for Abbey and from agribusiness and modern capitalism for Berry. Yet each shows a fascination with ruins, evidence of what Berry calls "the erasure of time" by which human marks on the landscape are slowly obliterated. This fascination reflects both confidence in the resilience of a natural world in relation to which human works seem transient and a need to see their own experience in relation to past human inter- actions with a place. Oliver has shown a sensitivity to the "mutila- tions" of the green world by European settlers and to more recent alterations of her familiar pinewoods, although we see relatively lit- tle attention to human influences in the natural world that she explores and celebrates in most of her poetry.

In their different ways all of the writers that I consider offer visions of an ideal nature, some with more awareness of the limita- tions of such visions than others. Audubon in his early ramblings discovered landscapes that appeared to him Edenic in their variety and abundance. He shared the tendency of William Bartram and other early travelers to celebrate New World versions of the earthly paradise. Later, he would lament the loss of other, "almost uninhab- ited" landscapes such as the Ohio valley as he first knew it. Thoreau's lost ideal was a "primitive" nature that he associated with the Native American predecessors of the farmers and villagers who had done so much to reconfigure the landscapes around Concord. Muir found his ideal nature in the Sierra range, seeing the terrain he explored in his early years in California as a wilderness still largely unspoiled although threatened by the incursions of miners and sheepherders and by the beginnings of tourism.

The imminent completion of Glen Canyon Dam shaped Abbey's sense of the canyon itself as a doomed Eden, an allusion he used for rhetorical effect, when he made the trip down the Col- orado River that he describes in *Desert Solitaire* (1968). Despite his concern with the advances of development and tourism, he contin- ued to find in desert landscapes a "world beyond." His ideal nature was a mysterious and ultimately unknowable world, the antithesis of the busy and circumscribed urban America from which he peri- odically retreated. Berry found his wilderness ideal in the Red River Gorge of Kentucky, marked by the passage of earlier human inhabitants but retaining enough of its original character, at least in places, to offer an experience of wilderness and to enable him to imagine the landscape as Daniel Boone might have known it. Forests, particularly a grove of large trees on his own farm to which he regularly retreats, serve Berry as a source of inspiration (especially for his "Sabbaths" series of poems) and an emblem of the kind of order and serenity he finds in the natural world. Oliver imagines the "green dazzling paradise" found by Meriwether Lewis as a lost ideal but for the most part finds the truth of the natural world in the local settings that she seeks out in her walks. These ordinary landscapes hold the promise of "earthly delights," even the possibility of moments in which a transcendent reality becomes visible.

These writers reveal an extraordinary desire for an intimate connection with the natural settings that they explore and in some cases inhabit. The kinds of intimacy that they seek vary, but all depend upon a capacity to "pay attention," as Oliver would put it. Paying attention implies approaching the natural world with an alertness and receptivity to the unexpected and the strange that make discovery possible. For most of these writers, it also implies an awareness of the difficulties of crossing over into the "other" world of nature. Berry enacts such a crossing in his poem "The Heron," showing himself leaving behind the labor and anxiety of a summer of farming to carry his boat down through the morning fog to the river, where he goes "easy and silent" in a world in which he becomes aware of warblers flashing through the trees and finds himself observed by the heron in its stillness: "Suddenly I know I have passed across / to a shore where I do not live."[26] Berry also knows that the river (the Kentucky) has changed, as a result of increased human use and abuse of the watershed, but in the poem he chooses to emphasize the continuities and the patience that he finds in the natural world, understood here as a world apart.

Intimacy with the natural world is often associated with a sense of wonder at the unexpected or the seemingly mysterious. Audubon frequently conveys such wonder upon coming across a bird he has been seeking; sometimes, as when he describes himself as enchanted with the scenery of the Mississippi or Ohio valleys, his responses seem to be shaped by the expectations of his audience for revelations of the marvelous. Thoreau reveals moments of wonder at the tumbling flight of a merlin or the pure, bright light of a spring morning. Muir can sustain a sense of wonder for pages of description of the pleasures of a "glorious Sierra day" of rambling through glacial landscapes and mountain meadows. With his deliberate abrasiveness and his frequent recourse to irony and sarcasm, Abbey might seem incapable of wonder, yet he can find the marvelous in the canyon of the Escalante and in moments when he yields to his sense of the silence and vastness of the desert. At such moments a perception of space appears to suspend his consciousness of time passing ("Light, space. Light and space without time").[27] Berry often associates wild nature with the unexpected. He can delight in the surprise of coming upon a quiet clearing or finding the floor of the woods covered with bluebells. In Oliver's poetry wonder takes the form of a capacity for amazement. She frequently shows herself "amazed" by unfolding natural dramas and describes herself as wanting to be able to say at the end: "all my life / I was a bride married to amazement."[28] Such amazement is her measure of truly living.

Wonder implies rapt attention by a passive observer: "[I] stood in my lonely body / amazed and full of attention."[29] Yet intimate contact with the natural world can also be energizing for the writers I describe. For one thing, it can lead to a charged sensuous awareness. Thoreau found himself restored by the wildness he found in the booming of the snipe and the smell of sedges in a marsh. Muir, clinging to the swaying top of a Douglas spruce in a Sierra windstorm, seems preternaturally sensitive to the effects of light on rippling trees, the mingled fragrances, and the symphonic music of the storm. The association of wild nature with health and vitality, apparent to some degree in all these writers, motivates explorations ranging from leisurely walks to rigorous explorations (in Maine, the Sierra, the deserts of the Southwest). A commitment to walking, usually understood as spiritual as well as physical exercise, links writers as disparate as Audubon and Oliver. Such walking, typically solitary, becomes a way of adapting oneself to nature's rhythms and experiencing its vitality.

It can also be a form of liberation. Thoreau sees walking as releasing him from the influence of European civilization, as well as from the conventions of village life: "westward I go free."[30] In wild nature, "not yet subdued to man," he finds a stimulus to freedom of thought and imagination.[31] The "freedom complete" that Muir experiences as he saunters through the high meadows of the Sierra, focused on his sensations, has more to do with a sense of transcending normal human limitations in the rapturous contemplation of a nature understood as instinct with divinity. Abbey, a river rafter as well as a walker in the desert, finds release from what he regards as the petty tyrannies of ordinary life in taking to the river. The "primeval liberty" that he experiences when he sets out on the Colorado has to do with his sense of escaping into an elemental world where he can relish what he perceives from moment to moment. For Abbey, this is among other things a liberty to flout authority of all kinds. Wilderness functions for him as a refuge from the kinds of order society imposes. For Berry, freedom typically involves a release from anxiety into a calm and authority associated with the natural world, as in "The Peace of Wild Things": "For a time / I rest in the grace of the world, and am free."[32] Such grace absolves him not from sin but from worry about the future. A sense of freedom from the dullness of conventional life is implicit in Oliver's description of herself as setting out each day "along / the green paths of the world."[33] For her, as for Thoreau, the most alive is the wildest. In fact, the linkage between freedom and wildness that Thoreau establishes at the beginning of "Walking" ("I wish to speak a word for Nature, for absolute freedom and wildness") is one that all these writers make in one way or another.[34]

By describing some of the motifs that connect the writers I discuss I do not mean to slight their differences or the fact that they reflect shifting cultural attitudes, with what we might regard as biases and omissions of various kinds. I hope that many of these differences and cultural shifts will become apparent in the chapters that follow. For example, Audubon's desire for intimacy with the birds he pursues means something quite different from the sense of intimacy Oliver creates in her evocations of encounters with deer. With Audubon, the urge to possess is paramount. His object was to fix his subjects, with as much fidelity as he could manage, through the mastery of his art. To do this he first had to shoot them, common practice for naturalists of the time. The admiration Audubon reveals for the birds he first observes in order to record their behavior yields to a form of conquest, driven by his ambition to produce the definitive rendering of the birds of America. Discovering and naming new species was another way of possessing them. Oliver, like Audubon, presents herself as an observer of nature's secrets. Yet she appears much less an intruder than Audubon, who always seems to be on the verge of disrupting the tranquillity of the scenes he observes; she is more an outsider yearning to share the instinctive life of creatures insofar as she can imagine this. Her concern is with preserving the memory of the intense moments in which she seems closest to connecting with them, as when the sight of a doe walking with her newborn fawn, "like a dream under the trees," prompts an overwhelming desire to begin life again and "to be utterly / wild."[35]

Juxtaposing any of the writers I consider, even those with such apparent affinities as Thoreau and Muir, has the effect of throwing their differences into relief. As Barry Lopez observes in *Arctic Dreams* (1986), we all apprehend the land imperfectly, with perceptions colored by preconception and desire.[36] Inevitably, we all apprehend it differently. Yet I would argue that the writers I consider here, along with many others, constitute a vital and still developing tradition of writing about wildness (and wilderness) in America. Despite their often striking differences, they reveal important commonalities in their desire to perceive and experience wildness and in the ways in which they conceive it. All assume a value inherent in the natural world and find contact with it energizing and illuminating. They may cultivate a sense of timelessness, through a focus on living intensely in the present. They may try to achieve a feeling of harmony with nature, even ecstatic or visionary moments. Such writers inevitably reflect an array of cultural influences, but their writing draws its strength from their actual encounters with a natural world that they see as exerting its own powerful influence and from the strategies they have discovered for representing these encounters and their implications. We continue to read them for the uniqueness and force of the visions generated by their experience of a nature they perceive as wild.

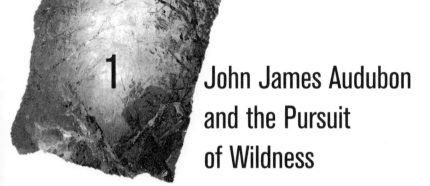

1 John James Audubon and the Pursuit of Wildness

> *Collected*
> *for my America its wildness, made*
> *distant grass & near Grebe*
> *to shine alike with detail, clear*
> *as water &*
>
> *find my old mind composed*
> *of what I saw*
> *& cannot be seen again,*
> *wilderness*
> *moves away from us steadily.*
>
> —PAMELA ALEXANDER,
> *Commonwealth of Wings*

In the introduction to the fifth and last volume of his *Ornithological Biography* (1839), published a year after the final plates of the *Birds of America* had appeared, John James Audubon represents himself as a "weary traveller" at last able to put aside his gun and knapsack and return home to rest and enjoy the pleasures of family. He relishes the triumphal moment, boasting that he has doubled the number of species "discovered, figured, and described" since the publication of *American Ornithology* (1809–29) by Alexander Wilson, the predecessor he had set out to surpass at a time when he had failed in business and had nothing to rely upon but confidence in his talent and a passion for wandering the woods in search of new birds to draw. Audubon celebrates the completion of his grand quest as a victory of perseverance: "In health and in sickness, in adversity and prosperity, in summer and winter, amidst the cheers of friends and the scowls of foes, I have depicted the Birds of America, and studied their habits as they roamed at large in their peculiar haunts."[1] Audubon had become adept at constructing a public image that would serve his purpose of selling the subscriptions by which he supported himself and his family (at a cost of just under one thousand dollars apiece to American subscribers). In expressing his relief and pride at the successful completion of his immense project, he was also consolidating the image of himself as heroic naturalist and artist who prevailed despite daunting odds.

Audubon makes his success seem more remarkable by confessing to potentially disabling fears and uncertainties. A catalogue of nightmares that startled him awake in the "wild woods" reveals a darker side of his imagination and also displays the tendency to exaggeration, to the point of melodrama, that characterizes Audubon's public writing. The apparitions that pursue him include sickness and poverty that almost caused him to abandon his great "task," threats from "Red Indians" and "white-skinned murderers," snakes that entwined him while "lean and ravenous" vultures looked on. The most concrete and compelling is the last: "Once, too, I dreamed, when asleep on a sand-bar on one of the Florida Keys, that a huge shark had me in his jaws, and was dragging me into the deep" (OB, 5:v). This alarming image suggests a fear of the monstrous, of nature as a devouring mouth, that threatens the sense of the cheerful tranquillity of the natural world that Audubon liked to project. He counters it with another revealing dream image that shows him transcending all the limitations and fears that haunt him. In this dream Audubon hears the songs of unknown birds, "beautiful and happy creatures," coming from a serene sky: "Then I would find myself furnished with large and powerful wings, and, cleaving the air like an eagle, I would fly off and by a few joyous bounds overtake the objects of my desire" (OB, 5:vi). The dream, which Audubon suggests was a recurrent one, expresses a vision of a paradisal nature (the air is "perfumed" and filled with "thousands of melodious notes") and one that he can enter and possess by assuming the power of the eagle. To achieve the kind of intimacy with wild nature that Audubon sought meant not only to observe "the objects of [his] desire" but to kill them (to "overtake" a bird usually meant to shoot it) in order to achieve the kind of accuracy he sought in his art. Modern readers are invariably struck by the paradox that Audubon had to destroy the birds he loved in order to create his enduring images of them.

In the introduction to the first volume of the *Ornithological Biography* Audubon offered a romantic view of himself as a child of nature, enjoying an untroubled "intimacy" with his surroundings and responding with delight to his father's instruction on birds and their habits. He represents himself as "[gazing] with ecstasy upon the pearly and shining eggs" in the nests he finds and then watching the development of the young birds (*OB,* 1:6). Audubon describes a progression from wanting to possess everything he sees to recognizing that the only way to preserve the "life" of the birds that he has begun to shoot is to "copy Nature." He describes his early, inadequate drawings as resembling "mangled corpses on a field of battle" (*OB,* 1:viii). The image suggests a consciousness that he has succeeded only in destroying the birds, not yet able to reanimate them through his art. The method that he eventually developed, and that he describes in part in his first introduction, reflects an extraordinary determination to make his representations of nature lifelike— by wiring freshly killed birds in natural postures to be drawn as quickly as possible; by taking exact measurements of the anatomical features of the birds (bill, legs, and so forth) in order to render their proportions accurately; and by insisting that the engravings made from his finished paintings be printed on double elephant paper, despite the objections of subscribers, to allow for life-size representations of even the largest birds. Audubon defended the sometimes surprising postures of his birds by appealing to his observations. He had seen swans drag a leg when swimming and herons drop their wings, as if dislocated, when warming themselves. Preserving the size and the proportions of his subjects, as well as the colors of their plumage, was essential to making them live on the page for Audubon, although the dramatic, even theatrical poses in which he often represents them may have more to do with our sense of their vitality than the scale and exactness of the images.[2]

In his first introduction Audubon describes his great pleasure in returning from his early morning rambles in Pennsylvania, where his father had sent him at eighteen to escape Napoleon's draft, "wet with dew, and bearing a feathered prize" (*OB*, 1:ix). This sense of the excitement of the chase and the satisfaction of obtaining the "prize" persists in his final introduction, although it is transferred here to the reader Audubon hoped to persuade to continue a work for which he saw himself as no longer physically fit. He was well aware, from the reports of Thomas Nuttall and other naturalists who had ventured into the West as Audubon had not yet managed to, that there were other species of American birds that remained to be described. In his exhortations to the reader, now imagined as disciple rather than subscriber, Audubon perpetuates the image of the solitary adventurer: "shoulder your gun, muster all your spirits, and start in search of the interesting unknown" (*OB*, 5:viii).[3] The romantic scene of discovery that he imagines, in a declivity of the Rockies with cliffs towering overhead, projects the role he has created for himself into the future and offers a revealing account of the actual process by which he transformed the "objects of [his] desire" from subject of scientific study to artist's model to delectable supper. We see his imagined successor shooting "a most splendid 'American pheasant,'" measuring and describing it, drawing it, skinning it and studying its organs, and finally cooking the "plump" bird on a fire of buffalo chips. The next day he rises with renewed vigor and desire, packs his new bird's skin, rolls his drawing around his previous ones, and pushes on, eventually returning to the settlements and publishing the results of the journey (*OB*, 5:x). Publishing the image of the bird, in the form of an engraving made from a painting based on the initial drawing, was the way Audubon finally took possession of it. Interestingly, he saw eating his subjects as the natural conclusion of his study. After rhapsodizing over the song of his beloved wood thrush, at one point, Audubon observes that its flesh is "extremely delicate and juicy" (*OB*, 1:374). The *Ornithological Biography* is among other things a gastronomic guide; Audubon ate his way through the birds of America.

In his later forays, undertaken after he had begun to publish the *Birds of America,* Audubon's role was actually that of leader of expeditions, typically supported by cutters provided by the agency of the Secretary of the Navy. The persona that he gives his imagined disciple resembles the one he invented when he first introduced himself to his predominantly British audience, that of the "American Woodsman." Audubon knew how to play on the growing romantic appeal of the American frontier to the European imagination. When he arrived in Liverpool in 1826, in a wolfskin coat and with his long hair groomed with bear grease, he could have been a stand-in for Daniel Boone, who had by then assumed the proportions of myth.[4] Audubon played the part of frontiersman when he arrived in London for the first time, telling stories and demonstrating bird calls and wolf howls at dinner parties, acting a role that he had polished in the travels that took him from Liverpool to Manchester and Edinburgh.[5]

The images of his solitary ramblings with dog and gun that Audubon presents in the *Ornithological Biography,* in the introductions and in the "episodes" describing frontier life that he included in the first three volumes, have a basis in fact but take on a strong romantic coloring. At one point he invites his readers to imagine him rising in the early dawn (by an "alder-fringed brook of some northern valley, or in the midst of some yet unexplored forest of the west") and serenely enjoying the melodies of the birds, then setting out invigorated. After lunch (of squirrel or trout) and rest, he wanders wherever his quarry may lead him, then builds a shed of boughs and enjoys an evening meal of "Widgeon or Blue-winged Teal, or perhaps the breast of Turkey, or a steak of venison." Thanking God for protection and "the sense of the Divine presence in this solitary place," he wraps himself in his blanket and falls into a sound sleep (*OB,* 4:vi–vii). Audubon represents himself here as content with his simple life and totally self-sufficient in the woods, offering an idealized vision of perfect harmony with the natural world, yet he characterizes himself as able to delight in such a life because his adventures have given "toughness" to his body and "elasticity" to his mind (*OB,* 4:vi). In the course of the *Ornithological Biography* he gives his readers enough glimpses of the physical and psychological hardships of the chase (the long trudges, the heat, the mosquitoes, and the potentially disabling fears and doubts) to make his quest seem heroic and to distance himself from the "closet-naturalists" he scorned. He creates the image of an American wilderness that offers idyllic moments, amid Edenic abundance, but only to someone with his stamina and resilience. Only to the true American Woodsman, in other words.

Yet Audubon had to be more than the American Woodsman to win the confidence of his intended audience. He actually created a dual persona in the *Ornithological Biography,* establishing himself not only as the American Woodsman, bringing back reports of adventures in the wilderness, but also as someone who could move in the social world of his moneyed subscribers. Audubon had to demonstrate that he was at home in both these worlds. He describes himself as writing about wild America from his comfortable quarters in Edinburgh, using his notes and journals from the field, with his young collaborator William Macgillivray "completing the scientific details and smoothing down the asperities" of his writing.[6] His introductions served Audubon primarily as a way of establishing a relationship with his readers that would ensure his credibility and sustain their interest in the enterprise and their trust in him. In them he offers advertisements for himself in the form of fresh reports from the field, as he journeys to such places as Labrador and Texas in pursuit of more species, and reports of subscribers gained and recognition achieved. Naming and praising notable friends and subscribers, and the growing circle of naturalists who provided help, constituted effective self-promotion, as well as acknowledgment. We hear of more and more prominent people in some way drawn into the effort, as Audubon's name-dropping charts his progress through the social and political elites he sought out in pursuing his project. Reporting that he had dined with President Andrew Jackson in the White House (on wild turkey shot in the environs of Washington) was a way for Audubon to confirm his emergence as a celebrity. Wilson, he notes, never made it to the White House.

Audubon's exaggerations and evasions complicate the task of understanding the "personal history" that he claims to present in his introductions. His most conspicuous evasions had to do with the facts of his illegitimate birth in Santo Domingo (now Haiti) to Jean Audubon and Jean Rabin, a serving maid from Nantes who had taken refuge on Audubon's father's plantation and who died soon after her son's birth.[7] In the introduction to the first volume of his *Ornithological Biography* Audubon describes himself, vaguely, as having "received life and light in the New World" (*OB,* 1:5) and represents his move to Pennsylvania as a return home, to "the woods of the New World," as if to establish his American identity. He sometimes told acquaintances that he was born in Louisiana, as the papers his father had supplied him with when he first came to America claimed.[8] Audubon showed a similar tendency to reconstruct his past in describing his father as having been at Valley Forge with George Washington (there is no evidence that he knew Washington at all) and in recounting a hunting expedition with Daniel Boone and Boone's tales of his own adventures (Boone in fact declined Audubon's invitation to hunt with him). Most biographers doubt that Audubon studied with Jacques Louis David, as he repeatedly claimed to have done.[9]

Audubon's talent for reinventing himself and embellishing his family history was encouraged, no doubt, by the circumstances of his birth and the disruptions of his early years. In France, where he was brought at the age of six after a slave uprising in Santo Domingo, he was formally adopted by Jean Audubon and his French wife and baptized as Jean Jacques Fougere Audubon. When he was growing up in Nantes, roaming the countryside and beginning to draw birds, he referred to himself for a time as "La Forest," a name that he continued to use with his wife, Lucy.[10] After being sent by his father to property he owned in Mill Grove, Pennsylvania, he assumed the name John James Audubon, which he regularly used, although he described his history variously to those he encountered. By the time Audubon presented himself to his public in the first volume of the *Ornithological Biography,* he had developed the persona of the American Woodsman and an idealized version of his childhood and his evolution as an artist. After receiving in England and Scotland the recognition that he had not found in Philadelphia or New York and discovering a suitable engraver in Robert Havell

Jr. of London, who was issuing installments of the *Birds of America,* he could speak with the assurance of someone who seemed to have put years of poverty and uncertainty firmly behind him. Audubon would continue to be evasive about his origins and self-conscious about his lack of scientific training, but he could write with justified confidence in his growing recognition when he set out to give his readers "some account of my life, and of the motives which have influenced me" (*OB*, 1:5).

In the text of the *Ornithological Biography* Audubon's capacity for exaggeration and invention appears primarily in the sixty episodes that he included in the first three volumes as a way of entertaining his readers. He drew upon his own adventures and observations of natural phenomena and frontier life for these but also adapted stories sent him by his friend John Bachman and others, typically representing them as drawn from his own experience. The episodes enabled Audubon to tell some lively stories that would appeal to those fascinated with the American frontier, stories that are sometimes humorous, often melodramatic, and full of picturesque details about frontier types and local customs.[11] Audubon himself appears as the solitary naturalist ("My knapsack, my gun, and my dog were all I had for baggage and company" ["The Prairie," *OB*, 1:8]), enduring physical hardship when necessary and welcoming the hospitality of a squatter's cabin. In addition to giving the reader some relief from the weight of fact in the bird biographies, the episodes allowed Audubon to show scenes from his life as naturalist and to offer the kinds of observations (about landscape, frontier customs, the power of nature as seen in floods and hurricanes) that reinforced his image as an expert on frontier America. My primary concern, however, is with the biographies themselves.

Audubon published his *Ornithological Biography* in five volumes, from 1831 to 1839, as a companion to the plates of the *Birds of America.* The sixty "episodes" that he included as a way of leavening his natural history have drawn more critical attention from those who have considered Audubon as a writer (not many) than the approximately five hundred accounts of bird behavior that constitute most of the text. These biographies, as Audubon thought of them, report details of migration, feeding, mating, and so forth, with a thoroughness unlikely to appeal to anyone but a hard-core birder, but they also offer revealing glimpses of Audubon's attitudes toward his subjects, his experiences in pursuing them, and his perceptions of a wild America in the process of being domesticated.

In his bird biographies, Audubon like other early naturalists saw himself as extending the project of naming and describing individual species, in the tradition of Linnaeus, but he showed an unusual interest in patterns of bird behavior, which he saw as manifesting the intricate order of the created world. His detailed descriptions of behavior, based upon his field notes and upon the reports of others when he lacked firsthand experience, constitute the heart of the biographies. Audubon was also unusual in establishing a strong sense of his personal presence. His tendency to theatricalize the birds that he painted has a counterpart in his dramatization of his efforts to find and observe them. Alexander Wilson comes across as a much less assertive and colorful narrator. Although Audubon lacked Wilson's assured command of English, having grown up speaking French, he wrote with a passion and a flair for self-dramatization that clearly distinguish him from his predecessor. Establishing a strong voice and elaborating on the difficulties he encountered was one way of asserting his authority in a field in which Audubon was conscious of operating as an amateur (he periodically faults Wilson, and sometimes William Bartram, for basing conclusions upon insufficient observation), and it was also a way of sharing with his readers his own sense of intimacy with the natural scene and the birds he pursued.[12]

Audubon frequently opens a biography anecdotally, drawing the reader into the scene of his observations. He begins one by insisting that he gets "a thousand times more pleasure" watching the purple gallinule "flirting its tail while gaily moving over the broad leaves of the water-lily, than I have ever done while silently sitting in the corner of a crowded apartment, gazing on the flutterings of gaudy fans and the wavings of flowing plumes" (OB, 4:37). The comparison is a revealing one, suggesting how much more alive and comfortable Audubon feels in the natural setting, although he is an observer in both. The anthropomorphism of the comparison is typical—Audubon often explains the habits of birds by appealing to human patterns of behavior—here serving to establish a telling contrast. The natural, instinctive movements of the gallinule are made to seem more genuine and engaging than the learned social behavior of the ladies, who, as Audubon was well aware, owed their feathers to the plume hunters of the New World.

Audubon places himself in the scene, lying concealed under a tree overlooking a Louisiana bayou when the male gallinule glides past looking for his mate: "Now he comes. . . . Look at his wings. . . . Now both birds walk along clinging to the stems and blades" (*OB*, 4:37). Such gazing offered a way of possessing the birds he pursued, at least for the moment, free from the social complications of the drawing room. The passage works by giving the reader the illusion of watching the action unfold in the presence of an enthusiastic and knowing guide.[13] Audubon's use of the present tense to establish a sense of immediacy, catching the reader up in the moment, is characteristic. He can be insistent in his effort to include the reader: "My light canoe is ready. Leap in, seat yourself snugly in the bow, and sit still while I paddle you to the green islands of this beautiful lake, where we shall probably find a Merganser or two" (*OB*, 4:263). In this instance, Audubon goes on to describe the sudden appearance of a swimming muskrat and of blooming dogwood and maples, "whose rich red blossoms cluster on the twigs," sketching the spring scene. Such detail has relatively little to do with the birds he will describe, but it evokes a world—wild, lush, distinctively American—that he would have his readers experience.

Macgillivray's revisions of Audubon's drafts tend to lower their emotional temperature, diluting the sense of excitement and the feeling of immediacy that they convey.[14] Audubon's own prose has a headlong quality, with erratic punctuation and frequent exclamations and addresses to the reader, some of which Macgillivray omits. By making changes in the service of clarity and gentility, Macgillivray loses some of Audubon's particularity, as well as some of the inflections of his voice. In the printed version of the biography of the "Great Northern Diver or Loon," for example, Macgillivray substitutes "active" for Audubon's "shy" in describing the loon and omits details about the location of the loon's nest near the water, as well as the fact that it flies not only fast but "to and from great distances." He also omits Audubon's injunction to the reader: "and then Gaze upon the bird *Reader* whilst it lays almost beneath your eyes, warming into life its dark coloured eggs."[15] By truncating and depersonalizing Audubon's description of shooting an anhinga, Macgillivray loses the tension with which Audubon invested the moment. The printed version reads: "Alas! it was not aware of its danger, but, after a few moments, during which I noted its curious motions, it fell dead into the water" (*OB*, 4:140). Audubon had written, "but alas! after a few moments spent in observing its graceful and yet strange movements, eagerness of possessing the bird would make my fingers touch a triger, and along with the instantaneous report of the piece, downward and quite dead I would see the [prize] fall to the Water."[16] Macgillivray's frequent changes, although often minor, have the effect of abridging Audubon's field observations and making his relationship to his reader less personal and emotionally charged. There is no evidence that Audubon objected to Macgillivray's relatively free editing; he appears to have welcomed his collaborator's help with the writing, as well as with the scientific descriptions, pressed as he was to finish the project and no doubt still insecure about his command of English prose.

Despite the effects of Macgillivray's editing, the sense of drama that Audubon gives many of the biographies survives in the printed version. He frequently begins a biography by recounting an adventure. One characteristic type involves a difficult approach to a sanctuary of some kind, where he can enjoy an unusually intimate view of the birds he is pursuing. In his account of the Florida cormorant, for example, Audubon describes a particular day in the Florida Keys (8 May 1832) on which he wades up a narrow channel with overarching mangroves and canes, in water up to his armpits, to discover twenty or more pairs of cormorants engaged in courtship behavior and mating:

> The males while swimming gracefully round the females, would raise their wing and tail, draw their head over their back, swell out their neck for an instant, and

with a quick thrust of the head utter a rough guttural note, not unlike the cry of a pig. The female at this moment would crouch as it were on the water, sinking into it, when her mate would sink over her until nothing more than his head was to be seen, and soon afterwards both sprung up and swam joyously round each other, croaking all the while. (*OB,* 3:390)

Audubon shows himself absorbed by the scene, refraining from shooting as he normally would have, and finally retreating, "almost exhausted with heat, my eyes aching from perspiration," when his advance causes the birds to dive or to fly away. The sense of strangeness and wonder with which he invests this scene is more striking for following another in which Audubon describes the "frightful havock" he and his party cause by firing into large breeding colonies of cormorants, "murders" he excuses as necessary for his work.[17] Audubon's extreme language and lurid detail (bodies floating on the water and heaped in the boat, crippled birds making for the open sea) suggest an effort to show himself conscious of the destructiveness of such indiscriminate shooting even as he participates in it. In fact, the party shoots far more birds, here and in comparable incidents, than he could use for his drawing. What he does not say is that birds were also shot for their skins, which could be sold to collectors. Audubon concludes his account of the carnage by switching to the scene of writing, "peaceably scratching my paper with an iron-pen, in one of the comfortable and quite cool houses" of Edinburgh (*OB,* 3:389), thus placing himself in the genteel world of his readers, recollecting in tranquillity the heat and emotion of his Florida adventures.

Audubon could describe a scene in which he blasts away with his companions followed by another in which he shows himself conscious of violating a private space with no apparent sense of incongruity, presenting the first as a scene of melodramatic horror and the second as a scene of discovery and moral choice (not to shoot). In both cases Audubon shows humans disrupting tranquil natural settings, more clearly in the manuscript than in the printed version. In passages omitted by Macgillivray Audubon describes the scene before the shooting starts as one in which the cormorants were "full of Life's greatest Joys until we entered their naturally peaceful retreats" and the place where he discovered them mating as a spot "sacred to these birds."[18] He begins another biography, that of the common cormorant, by describing himself crawling out on a precipice above the Labrador cliffs, on which he found the bird nesting, to observe a mother fondling and feeding her young, "quite unconscious of my being near" (*OB*, 3:458). He fixes the time (3:00 A.M. on 3 July 1833) and describes the gale that dashes the sea against the rocks below, masking his approach. The scene ends abruptly with Audubon crawling back from his perch after the mother has discovered him and taken flight in alarm and the young have crawled into a recess in the cliff, satisfied that he has observed both their pleasures and their terrors. As in the case of the Florida cormorants, Audubon portrays himself here as an intruder who violates the sanctuary he is so eager to observe, unable to keep himself from pressing beyond the point at which he will trigger flight. In each case the spell of the moment is broken, and Audubon retreats, implicitly justifying his disruptive intrusion by the knowledge (of the details of nurturing or courtship) that he can share with his readers.

By describing his foray on the cliffs of Labrador Audubon could impress the reader with his passion and his daring as a field naturalist. A visual image that he included as part of the background of his painting of a golden eagle gives perhaps the best indication of the sense of himself as romantic adventurer that he sought to convey. This image, omitted in Havell's engraving, shows the small figure of a hunter with a gun and a large bird slung on his back pulling himself across an abyss by means of a log that he straddles, with rugged mountains as a backdrop.[19] Ironically, Audubon drew his golden eagle from a captured specimen supplied by a Boston acquaintance. He gives a semicomic account of an unsuccessful attempt to suffocate it with a charcoal fire that nearly drives him out of the house but has no effect on the bird (*OB*, 2:465).

Audubon recreates other moments that exemplify a deeper kind of intimacy with the birds he is pursuing, sometimes in wild settings in which the welcome sound of birdsong restores a sense of calm and well-being. He describes himself as "wearied, hungry, drenched" after a violent storm in the woods, questioning why he is there and whether he will see his family again, when he is cheered and consoled by the song of the wood thrush (*OB,* 1:373–74); he is similarly restored by hearing the song of the winter wren "from the dark depths of the unwholesome swamp" and inspired with "a feeling of wonder and delight" (*OB,* 4:431). Audubon characteristically pronounces such moments signs of the benevolence of God in saving him from despair, shifting the focus from the marvelous qualities of the song to his own state of mind. They suggest a fundamental optimism about the restorative properties of the natural world, seen as part of the divine plan, and a confidence in his own ability to achieve a state of harmony with his wild surroundings. He portrays the thrush's song as dissolving his anxieties and making him feel once more at home in the woods.

In reading Audubon's bird biographies one often feels a tension between his absorption in the world of the birds he is observing, as he delights in their song or in the intricacies of their behavior, and his use of language that has the effect of assimilating them to his own. He has been criticized for anthropomorphism, in his paintings as well as his prose. In the prose this appears as a tendency to see patterns of human social behavior and to render moral judgments, implicitly or explicitly. Thus the blue jay becomes tyrannical and thieving, and the pigeon hawk devours his prey and returns for more, "bent on foul deeds" (*OB,* 1:466). Audubon regularly describes the period of mating as the "love season" and attributes human qualities to birds engaged in courtship behavior. He begins his account of the Carolina turtledove with the explanation that he has painted the doves on a branch that has "a profusion of white blossoms, emblematic of purity and chastity" (*OB,* 1:91), then presents a "love scene" in which the female coyly delays the gratification of her "lover" with "virgin-like resolves to put his sincerity to the test." The subsequent description of the movement and habits of the doves avoids this kind of humanizing of his subjects, but Audubon frequently engages in it, perhaps as a way of reaching his readers by creating a kind of drama that they could recognize. Amy R. W. Meyers has argued that Audubon "often constructs his pictures and his text as moral tales in the tradition of the fable," influenced by his fondness for the fables of Lafontaine.[20] Such fables offer a commentary on human as well as bird behavior, in the case of the doves suggesting the rewards of true courtship. Moralizing in this fashion was one way of enhancing the significance of bird behavior for his readers. Audubon's anthropomorphizing can also be taken as a measure of his own sympathy and of his desire to identify with his subjects and thus as another way of achieving the sense of intimacy that he sought.[21]

Audubon's frequently remarked use of florid language in his descriptive writing—evident in his drafts, as well as in Macgillivray's revised versions—reflects the tendency to idealize and theatricalize his subjects that underlies his anthropomorphizing habits. He seems unable to resist a superlative or an adjective that will heighten the drama of his own situation (mosquitoes become "blood-sucking," "tormenting") or the appeal of the birds he observes. Despite such excesses, however, Audubon often conveys precise detail in a relatively unembellished fashion. He will describe the way a grackle changes color in the light (*OB*, 1:35) or the look of white pelicans in flight: "when flying in flocks in clear sunny weather, the blue of their wings glistens like polished steel" (*OB*, 4:113). He may begin by creating a sentimentalized scene, showing himself lured out by "the clear light of the silvery moon" to the banks of the Ohio ("*la belle riviere*") to "indulge in the contemplation of nature," then drop this self-consciously literary manner as he records the activity of the mergansers that absorb him: "washing their bodies by short plunges, and splashing up the water about them," pluming themselves and "now and then emitting a low grunting note of pleasure" (*OB*, 3:246). In rendering another Ohio River scene Audubon rhapsodizes on the sunset but reveals a sharp eye in his description of the white pelicans that constitute his real subject: "Ranged along the margins of the sand-bar, in broken array, stand a hundred heavy-bodied Pelicans" (*OB*, 4:89).

Some of his most successful descriptive writing combines exact observation with a sense of wonder or delight that seems relatively free of mannerisms, as in his account of the great gatherings of trumpeter swans he observed when in winter camp on the Tawapatee Bottom of the Mississippi. Audubon describes how at twilight "the loud-sounding notes of hundreds of Trumpeters would burst on the ear; and as I gazed over the ice-bound river, flocks after flocks would be seen coming from afar and in various directions, and alighting about the middle of the stream opposite our encampment" (*OB*, 4:538). When wolves howled at night, "the clanging cries of the Swans would fill the air." On a sunny morning, "whole flocks would rise on their feet, trim their plumage, and as they started with wings extended, as if, racing in rivalry, the pattering of their feet would come on the ear like the noise of great muffled drums, accompanied by the loud and clear sounds of their voices." Audubon's description works here because he effectively conveys a sense of the wildness and strangeness of the scene, which for the most part resists assimilation to familiar patterns.

Audubon's accounts of the flight of birds often blend careful observation with an aesthetic delight in the grace of their movements. His capacity for empathy with his subjects is particularly apparent in such descriptions—for example, that of the flight of the immense flocks of passenger pigeons. He marvels at their vast numbers but also at their speed and noise and the intricacy of their movements, finding metaphoric language to charge his description with a sense of wonder:

> I cannot describe to you the extreme beauty of their aerial evolutions, when a Hawk chanced to press upon the rear of a flock. At once, like a torrent, and with a noise like thunder, they rushed into a compact mass, pressing upon each other towards the centre. In these almost solid masses, they darted forward in undulating and angular lines, descended and swept close over the earth with inconceivable velocity, mounted perpendicularly so as to resemble a vast column, and, when high, were seen wheeling and twisting within their continued lines, which then resembled the coils of a gigantic serpent. (*OB*, 1:321)

The surprising way successive flocks exactly replicate the evasive movements of the one attacked by the hawk heightens Audubon's wonder at what he presents as a spectacular and mysterious natural phenomenon. This mood yields to one of mesmerized horror at the devastation of the roosting flocks he goes on to describe, a scene of "uproar and confusion."

In describing birds in their habitats and recounting his adventures in pursuing them, Audubon offered among other things a vision of the American wilderness as a place of Edenic abundance and sublime spectacle. As I have suggested, Audubon established the sense of himself as resilient American Woodsman by insisting upon the hardships of wilderness travel. He tells his readers, for example, about making his way through a swamp watching for poisonous snakes and through "the most gloomy part of a thick and tangled wood" (*OB*, 4:431), enduring extremes of heat and cold and at times overcome by loneliness and a sense of failure. By dramatizing his physical and psychological ordeals, Audubon suggests the resistant and threatening aspect of wilderness and at the same time ennobles his own adventures. Yet his accounts of these ordeals also have the effect of heightening, by contrast, scenes that impress him by their natural beauty and sensuous delight or their compelling strangeness.

Audubon begins his biography of the mockingbird with an awed description of its lush habitat in Louisiana, where he claims that "the bounties of nature are in the greatest perfection":

> It is where the Great Magnolia shoots up its majestic trunk, crowned with evergreen leaves, and decorated with a thousand beautiful flowers, that perfume the air around; where the forests and fields are adorned with blossoms of every hue; where the Golden Orange ornaments the gardens and groves; where Bignonias of various kinds interlace their climbing stems around the White-flowered Stuartia, and mounting still higher, cover the summits of the lofty trees around, accompanied with innumerable Vines, that here and there festoon the dense foliage of the magnificent woods, lending to the vernal breeze a slight portion of the perfume of their clustered flowers; where a genial warmth seldom

forsakes the atmosphere; where berries and fruits of all descriptions are met with at every step;——in a word, it is where Nature seems to have paused, as she passed over the Earth, and opening her stores, to have strewed with unsparing hand the diversified seeds from which have sprung all the beautiful and splendid forms which I should in vain attempt to describe, that the Mocking Bird should have fixed its abode, there only that its wondrous song should be heard. (*OB,* 1:108)

Audubon's Louisiana could be a New World version of the earthly paradise of literary tradition, most memorably rendered by Milton in *Paradise Lost,* with its fragrance, vernal breeze, simultaneous flowers and fruit, and sense of containing all possible natural delights ("blossoms of every hue"). Yet Audubon's eye orders what others might perceive as rank profusion, as he describes flowering vines ornamenting the trees and suggests a pleasing alternation of forest and field, garden and grove. He makes the landscape seem an inviting one that we could walk through with ease, meeting natural wonders "at every step," rather than a forbidding tangle of vines and undergrowth. Audubon, unlike Milton, grounded his description in actual observation, but this was clearly conditioned by assumptions about the ideal landscape. Audubon may have been influenced by William Bartram's descriptions of the paradisal luxuriance he found in Florida, presented as "delightful," "Elysian," "sublime," and as a continuing source of amazement. In his *Travels* (1791) Bartram describes himself as tranquilly meditating on "the marvellous scenes of primitive nature, as yet unmodified by the hand of man."[22] Chateaubriand's popular romance *Atala* (1802), which Audubon could easily have known, had described Louisiana in similar if more extravagant terms as a scene of primal nature, a "New Eden" in which "trees of every form and every colour and every odour mingle," bound together by wild vines and towered over by magnolias with their "great white blossoms."[23]

In such scenes as the one in which he places the mockingbird Audubon pictures a marvelous, unspoiled nature that wholly delights the senses and invites a mood of rapt wonder; he describes the traveler as unable to ascend the Mississippi "without feeling enchanted" by the vegetation on its shores, then goes on to describe the "still more enchanting Ohio" (*OB*, 1:67). Chateaubriand and Bartram had established the expectation that enchantment was the appropriate response to scenes of wild nature.[24] Stephen Greenblatt sees expressions of wonder in early modern accounts of the New World as a form of appropriation, what he calls "a record of the colonizing of the marvelous."[25] I would see the wonder that I describe in Audubon as manifesting another kind of human yearning—to experience an alluring and seemingly pristine natural world that suggests the splendor of creation.

Particular scenes are animated and made enchanting for Audubon by whatever bird is the object of his pursuit, especially by the harmonizing power of its song, "Nature's own music" (*OB*, 1:109). In a description of the dense vegetation on the banks of the Mississippi, presented as the habitat of the Baltimore oriole, Audubon describes the delight of hearing the "clear mellow notes" of the oriole in such a solitary place. For him such song "insensibly leads the mind . . . first to the contemplation of the wonders of nature and then to that of the Great Creator himself" (*OB*, 1:67). Extended descriptions of habitats become means of celebrating these wonders and thus the divine design they reveal. Audubon typically associated birds with their natural surroundings in his paintings by presenting them against a background of vegetation and sometimes a landscape that would suggest their habitat.[26] He describes finding Baltimore orioles on the Ohio moving gracefully among "the pendulous branches of the lofty Tulip-trees" (*OB*, 1:67); one of his paintings shows orioles with their hanging nest in such pendulous branches, with the presence of the tulip tree itself asserted by sharply drawn leaves and a prominent white blossom.

Audubon began the first biography of his second volume, published in 1834, by proclaiming the wonders of the American wilderness in its various aspects, anticipating the more sustained case for the beauty and sublimity of America's wild landscapes that Thomas Cole would make in his famous "Essay on American Scenery"

(1836). Audubon characteristically presents his overview as a regis-ter of his own remarkable experience, embellishing this experience in the process:

> Amid the tall grass of the far-extended prairies of the West, in the solemn forests of the North, on the heights of the midland mountains, by the shores of the bound-less ocean, and on the bosom of the vast lakes and magnificent rivers, have I sought to search out the things which have been hidden since the creation of this won-drous world, or seen only by the naked Indian, who has, for unknown ages, dwelt in the gorgeous but melan-choly wilderness. (*OB*, 2:1)

What stranger, Audubon goes on to ask, "can form an adequate con-ception of the extent of its primeval woods . . . of the vast bays of our Atlantic coasts . . . of our ocean-lakes, our mighty rivers, our thundering cataracts, our majestic mountains, rearing their snowy heads into the calm regions of the clear cold sky?" (*OB*, 2:2) Such mingling of nationalism and the sublime would take hold in nine-teenth-century America, particularly among those seeking to invest landscape painting with a new significance.[27] Audubon reinforced his own identity as an American (he speaks of "my native land") by representing an idealized national landscape, at the same time enhancing the romantic appeal of the American wilderness for his European audience and establishing himself as a guide to its unknown splendors.

Audubon's expedition to Labrador in 1833 tested his enthusi-asm for the American sublime. He was excited by the vast concen-trations of seabirds and the spectacle of cliffs, mountains, and mossy valleys. And by scenes that did not conform to any sense he had of an ordered landscape: "The country, so wild and grand, is of itself enough to interest anyone in its wonderful dreariness. Its mossy, gray-clothed rocks, heaped and thrown together as if by chance, in the most fantastical groups imaginable." Yet such wildness could appear threatening, defeating efforts to assimilate it: "This after-noon I thought the country looked more terrifyingly wild than ever; the dark clouds, casting their shadows on the stupendous mosses of rugged rock, lead the imagination into regions impossible to describe." As the *Ripley* proceeded north, Audubon's sense of the oppressive desolation of the landscape increased, along with his fatigue from struggling to complete drawings under dauntingly cold and wet conditions. Looking from a high rock over "the most extensive and dreariest wilderness" he had ever seen, he could observe that "it chilled the heart to gaze on these barren lands of Labrador."[28] By the time the *Ripley* had returned to Newfoundland and the festive scene he and his companions found there ("A Ball in Newfoundland"), Audubon was clearly ready for a more humanized landscape.[29] He could find the rocky shores of Newfoundland "bold and magnificent," an appealing kind of "romantic scenery," because he saw a human presence there, as well as more luxuriant growth. The absence of these in Labrador and the "deep silence" he found there (*OB*, 3:420) have much to do with the melancholy sense of dreariness the landscape aroused in him. When the *Ripley* sailed into the Bay of St. George on a mild and sunny day, Audubon hailed the "welcome sight of cattle feeding in cultivated meadows, and of people at their avocations" (*OB*, 2:211). His encounter with a landscape as "barren" as he perceived Labrador's to be taxed Audubon mentally as well as physically; at this point in his wanderings, at least, he preferred his wilderness softened by the proximity of human settlement.

One of the paradoxes that strikes a reader of Audubon today is that he could evoke the splendors of unspoiled American landscapes and simultaneously embrace the progress of settlement. He follows the description of the Louisiana habitat of the mockingbird, with its dense growth of vines and flowering trees, by describing this paradisal place as found "in that great continent to whose distant shores Europe has sent forth her adventurous sons, to wrest for themselves a habitation from the wild inhabitants of the forest, and to convert the neglected soil into fields of exuberant fertility" (*OB*, 1:108). The assumption that such fertility should serve human needs, rather than be "neglected," is even more pronounced in Audubon's narrative of the settlement of Kentucky, which focuses on the heroic exploits of the settlers, led by Daniel Boone, in "forcing their way through the pathless woods to the land of abundance" ("Kentucky Sports," *OB,* 1:291). Audubon's romance of settlement assumes "the mental superiority and indomitable courage of the white men" who eventually displace the native inhabitants and represents the land itself as inviting exploitation, with "the richness of its soil, its magnificent forests, its numberless navigable streams, its salt springs and licks, its saltpetre caves, its coal strata, and the vast herds of buffaloes and deer that browsed on its hills and amidst its charming valleys" (*OB,* 1:290). The Kentucky Audubon knew was already considerably changed from its state prior to European settlement. In another episode he describes passing a large buffalo lick and asks, rhetorically, "where now are the bulls which erst scraped the earth away?" (*OB,* 3:379) It is not surprising to find Audubon appealing to the appetite of his audience for stories of pioneer exploits (he goes on to describe the extraordinary feats of marksmanship that constitute Kentucky "sports"), but it is striking that the episode shows none of the sense of loss that colors his nostalgic account of his own early trip down the Ohio with his family to settle in Henderson, Kentucky ("The Ohio"). In his story of the original settling of Kentucky the banks of the Ohio, "that magnificent stream," appear simply as inviting land to be cleared.

Whatever Audubon's efforts to portray the American wilderness, in its forbidding as well as its delightful aspects, he wrote from the perspective of someone who understood and for the most part accepted the interests of the settled world to which he belonged. While Audubon shows some birds serving the interests of settlers (the raven destroys pests and insects), he represents others as invaders of domesticated spaces, including the hawks and owls that prey on barnyard chickens and the rice buntings (bobolinks) and other species that descend in dense flocks to strip young grain from the fields. He begins his biography of the great horned owl with a minidrama, shown from the perspective of a passing boatman, in which the owl disrupts the "placid serenity" of a moonlit summer night by gliding down to attack a squatter's hen. The squatter, aroused by the cacklings of his hens, emerges with his rifle to shoot the "murderous owl" and restore the tranquillity of the evening (*OB,*

1:313–14). Audubon goes on to describe the habits of the owl in his usual fashion, but his emphasis is on the "havock" visited upon farms on the frontier by "these dangerous and powerful marauders" (OB, 3:316). In this case the "interrupted idyll," to adapt Leo Marx's term, is one in which wild nature disrupts a peaceful domestic scene.[30]

Audubon describes the "devastations" of the bobolinks from the perspective of the farmer and the "sport" of shooting them with no noticeable thinning of the flocks: "Millions of these birds are destroyed, and yet millions remain" (OB, 1:285–86) to move on to the rice plantations of the Carolinas. He leaves no doubt that the fields should be defended and the birds that are shot enjoyed as a delicacy: "Their flesh is extremely tender and juicy. The markets are amply supplied, and the epicures have a glorious time of it" (OB, 1:286). Audubon's frequent comments on the tastiness of birds reflect a residual sense that the creatures are intended for human use, as well as a lively interest in the state of the market, presumably shared by many of his readers (he questions whether canvasback ducks are worth the two dollars a pair they have come to command in New Orleans).[31]

Audubon's account of the Carolina parakeet, now extinct, reflects the unresolved tension between aesthetic appreciation and practical concern that one finds in his writing. He recognizes the destructiveness of the flocks that swarm to grain fields and also devour unripe fruit of virtually any kind, implicitly approving the "great slaughter" of them planters make for these "outrages" (OB, 1:136). Yet he admires the parakeets in what he sees as their natural habitat, finding that "the richness of their plumage, their beautiful mode of flight, and even their screams" have their charms in the "darkest forests and most sequestered swamps" (OB, 1:138). His observation that they are very rapidly declining in number and contracting their range implies a sense of loss, although he draws no conclusions about the effects of settlement on their habitat or feeding patterns. His painting of seven parakeets in an array of postures feeding on cockleburs suggests their abundance, as well as their animation and their brilliant coloring.

Audubon's comments about taming parakeets (by repeatedly

immersing them in water) and their suitability as companions (poor, given their disagreeable screams and the impossibility of training them to speak) suggest another common use of wild birds—as pets. Audubon's accounts of his and his friend John Bachman's experiments in taming a startling variety of birds (including herons and cranes) reflect a blurring of distinctions between wild and domestic that may seem surprising to modern readers accustomed to a narrower sense of the domestic. Such experiments might begin out of scientific curiosity and become something else. Audubon describes keeping a trumpeter swan for two years as a household pet for the entertainment of his family. He thought nothing of carrying a screech owl in his coat pocket on the train from New York to Philadelphia.

The impact of humans on the natural environment is most apparent in Audubon's accounts of hunting practices. Audubon shot birds to draw them and sometimes to collect their skins, but he also appears to have enjoyed the kind of "sport" shooting he frequently describes for readers given to the practice themselves. In fact, the *Ornithological Biography* sometimes reads like a precursor to *Field and Stream,* with its many hunting stories and practical observations. Audubon offers what amounts to a guide to hunting and fishing practices, the latter in episodes that detail methods used to catch perch and catfish in the Ohio. In describing the practice of firing into the huge flocks of red-winged blackbirds that invade grain fields, Audubon reports that "I have myself shot hundreds in the course of an afternoon, killing from ten to fifteen at every discharge" (*OB,* 1:350). Estimates of numbers killed, by Audubon himself or by those he had observed or heard about, served as a convenient measure of prowess with rifle and shotgun at a time when vast flocks were commonplace and no one worried about limiting the number of birds shot.

Audubon begins his biography of the woodcock by describing his anxiety at watching a mother bird trying in her "innocent simplicity" to preserve her brood from a pack of boys by feigning injury, then shows himself emerging from his thicket to save her. His account of the habits of the woodcock reveals a typically well-informed sense of characteristic behavior and an admiration for its skill at evading detection, but it also functions as a primer for the "thousands" who, like him, "are fond of Woodcock shooting" (*OB,* 3:478), with advice about where to find the birds and how to learn to shoot them. He ends his narrative with a wistful evocation of the pleasures of returning home from the hunt with a bag of wood-cocks, "fatigued and covered with mud," to enjoy a cheerful dinner with his family. Writing in Edinburgh, without hope of a woodcock dinner, as he reminds the reader, he conveys a palpable longing to be at the table with "a jug of sparkling Newark cider" and a mess of woodcocks arrayed on the platter: "so white, so tender, and so beau-tifully surrounded by savoury juice" (*OB,* 3:480). The movement of the biography from compassion for the "innocent" mother bird to sympathetic description of the habits of the woodcock to a fantasy of a woodcock dinner was natural for Audubon. Writing a little over a hundred years later, Aldo Leopold would describe how his obser-vation of the woodcock's courtship flight, also described by Audubon, moderated his hunting instinct.[32]

Audubon's descriptions of hunting practices can be more crit-ical. His biography of the clapper rail offers an account of the "exceedingly pleasant" sport of shooting rails on the South Carolina coast during rising tides that turns into a scene of melodramatic horror, inviting sympathy for the apparently helpless birds unable to escape gunners in the tidal creeks. Reacting to what clearly struck him as excess, Audubon invites the reader's sympathy for the rails, more simply and convincingly in his draft than in Macgillivray's embellished rewriting: "look upon them with pity as I do Reader, is it not a sorrowful tale, to think here after of the last gasping breath, of the last movement of their leggs, as of the declining colouring of their just now bright eyes."[33] The hunter becomes a "Cruel Sports-man," "besmeared by the black powder," who "laughs and makes merry" as he takes home three times as much game as his family can use.[34]

Seemingly contradictory attitudes toward hunting coexist in the *Ornithological Biography.* Audubon could sentimentalize the plight of the rail and still regularly report feats of marksmanship, even in this case ("One of them I have seen shoot fifty Marsh-Hens at fifty successive shots" [*OB,* 3:38]). Various episodes (e.g., "Deer

Hunting," "The Cougar," "Kentucky Sports") describe frontier skills
of stalking and shooting that he clearly admired and sought to render for an interested audience, with characteristic inventiveness in
his description of Daniel Boone's marksmanship in "Kentucky
Sports" (*OB*, 1:293–94). At times Audubon shows off his own skill.
He reports shooting single puffins on the wing "for one hour by my
watch": "How many puffins I killed in that time I take the liberty of
leaving you to guess" (*OB*, 3:107). The practices at which Audubon
recoils are those in which the destruction seems to him egregious
or without apparent point. His lurid account of the "devastation" of
flocks of roosting passenger pigeons depicts a nightlong "uproar" of
men with guns and poles, wolves and other predators, and hogs
loosed to feed on the remnants. In his description of the breeding of
the pigeons Audubon pictures "the tyrant of the creation, man" disturbing a tranquil scene by chopping down nest trees to get at the
squabs (1:326). Interestingly, he questions the assumption that such
practices would destroy the species, concluding that only the gradual reduction of the forests would diminish the passenger pigeon's
extraordinary numbers.

Audubon did display increasing concern about the decline of
wilderness and of the kind of abundance that was commonplace
when he began his wanderings. Sometimes this takes the form of
observations about the decline of a particular species such as the
Carolina parakeet. At other times it becomes a more generalized
fatalism. Audubon follows an account of vast flocks of migrating
birds rising from river banks at the approach of a pigeon hawk, so
dense that even the most inept "sportsmen" can hit several birds
with one shot, with a passage in which he indulges a sense of
inevitable loss. An awareness of how rapidly the country is changing
colors his recollection of the spectacle and heightens its drama for
his readers:

> When the Reed-birds, the Redwings, and Soras, shall
> have become so scarce as to be searched for with the
> same interest as our little Partridges already are; when
> the margins of our rivers shall have been drained and
> ploughed to the very tide mark; when the Grouse shall

have to be protected by game-laws; when Turkeys shall no longer be met with in the wild state;—how strange will the tale which I now tell sound in the ears of those who may walk along the banks of these rivers, and over the fields which have occupied the place of these marshes! (*OB*, 1:466–67)

Audubon laments the prospective loss of such marshes and the birds that depend upon this habitat, but with a resigned acceptance of his role as a teller of tales of a vanishing world.

Audubon's concern took a more urgent and practical form when he saw destruction on a much larger scale on his journey to Labrador (1833) and on his final expedition up the Missouri River to Fort Union (1843). The most egregious villains of the *Ornithological Biography* are "The Eggers of Labrador," "vile thieves" who trample nests as they strip islands of eggs and shoot guillemots indiscriminately. Audubon represents them as entirely depraved, given to violence and drunkenness as well as to brutality toward the colonies of birds on which they prey. In this case, unlike that of the passenger pigeon, he perceives the practices of humans as threatening the survival of a species, and he welcomes the attempts of the British government to check this "war of extermination" (*OB*, 3:85). In the journal he kept on his trip up the Missouri Audubon deplores the slaughter of the buffalo ("thousands multiplied by thousands of Buffaloes are murdered in senseless play, and their enormous carcasses are suffered to be the prey of the Wolf, the Raven and the Buzzard"), predicting their extinction: "before many years the Buffalo, like the Great Auk, will have disappeared; surely this should not be permitted."[35] Alton Lindsey sees "an older, more sensitive, more philosophical Audubon" in his later writings, particularly in the *Missouri River Journals*.[36] At one point Audubon laments, "Where can I go now, and visit Nature undisturbed?"[37] The older, more reflective Audubon shows more restraint in his collecting, Lindsey argues, yet given his heightened sense of a vanishing wilderness, it is surprising that Audubon did not show a greater will to check the practices that were contributing to its decline.

There is in fact a curious disjunction in the *Missouri River Jour-*

nals between Audubon's laments over acts of destruction he sees as threatening whole species and the practices he does tolerate. If he has lost some of the trigger itch apparent in his accounts of earlier adventures, he continues to shoot (larks, wolves, buffalo) and shows no concern over his younger companions' habit of shooting buffalo whenever the opportunity presents itself. He comments on how easy it is to shoot wolves from the walls of Fort Union in the evening, a diversion he may well have enjoyed himself. In Labrador Audubon deplored the piratical behavior of the eggers but tried to buy eggs and specimens from the only two he actually met and complained when they reneged on their promises. The heroes of this episode are the "hundred honest fishermen" who battle the eggers in the course of their efforts to procure eggs to vary their diet. Audubon describes breakfasting on ibis eggs himself in the Florida Keys (*OB*, 2:345).

One can differentiate the practice of gathering wild birds' eggs for personal consumption and the massive pillaging engaged in by the eggers, of course, but it is significant that Audubon's instinct for preservation was aroused only by extreme cases of abuse, and not always then. He reports the practices of the turtlers of Florida, including his pilot's claim that one man caught eight hundred green turtles in the course of a year, with no sense of concern about the effects of such practices. He is simply appealing to the curiosity of his readers about the exotic, not neglecting information about the market value of the turtles: they bring "four to six cents per pound" in Key West (*OB*, 2:376). Even when Audubon shows an awareness of the waste or destructiveness of particular practices, he does not always condemn them. He begins his episode on the live-oakers of Florida by describing a traveler, presumably himself, enjoying the fragrant woods of a hummock and being interrupted by the arrival of men with axes who bring down young trees with the mature ones and leave many of the latter on the ground because they prove to have rot. Yet the focus of the episode is on the habits and the skill of the live-oakers, for whom Audubon shows considerably more sympathy than he does for the government agent sent to protect the live oaks, revealing the agent to be so inept that he can't tell a live oak from a swamp oak.

Audubon's sympathy for squatters of various kinds, whose hospitality he frequently enjoyed and describes in a number of his episodes, helps to explain how he could celebrate the relatively undisturbed wilderness he knew from his rambles and at the same time accept the incursions made on it by settlers. His account entitled "The Squatters of the Mississippi" shows them leaving their worn-out land in Virginia and finally stopping their westward journey when they encounter the "dark deep woods" on the banks of the river. The story he tells is one of industry and accompanying prosperity, as the squatters progressively expand their logging activity and graduate from a "single cabin" in the woods to a "neat village":

> Thus are the vast frontiers of our country peopled, and thus does civilization, year after year, extend over the western wilds. Time will no doubt be, when the great valley of the Mississippi, still covered with primeval forests, interspersed with swamps, will smile with corn-fields and orchards, while crowded cities will rise at intervals along its banks, and enlightened nations will rejoice in the bounties of Providence. (*OB*, 2:134)

Such statements reflect the prevalent view that wilderness was a stage in the emergence of the new nation, the beginning point of a narrative of progress rather than a state that one could hope to preserve to any significant degree.[38] To question this story would have undermined the Americanism that Audubon was eager to demonstrate.

Yet one can see the other side of Audubon's assumption of the perspective of the settler in the elegiac strain that surfaces periodically in the *Ornithological Biography,* most memorably in the first of his episodes, "The Ohio." This episode can be read as an exercise in nostalgia—for the adventure of Audubon's trip down the Ohio to Henderson with his wife and young son twenty years before and, especially, for the pleasures of absorbing a still relatively undisturbed natural environment: "Leisurely we moved along, gazing all day on the grandeur and beauty of the wild scenery around us" (*OB*, 1:29). Audubon presents an October idyll of warm days and brilliant foliage that suggests a suspension of time. What gives the episode its drama, apart from a comic anecdote about the "uproar" produced by a Methodist camp meeting, is an undertone of anxiety about the impending change foreshadowed by traffic on the river and scattered evidence of settlement on its banks. In an abrupt shift of focus to the Ohio of the present near the end of the episode Audubon emphasizes the cost of progress, in the loss of "the vast herds of elks, deer and buffaloes which once pastured in these hills and in these valleys" and the continual "din of hammers and machinery" that replaces the comparative quiet of the former scene. Here Audubon idealizes the "state of nature" that the pioneers disrupted, embodied by forests he recalls overhanging the river, "still unmolested by the axe of the settler" (*OB,* 1:31).

The dissonance between the elegiac mood Audubon indulges here and his praise of heroic and enterprising settlers in other places has been read as ambivalence but is better understood as a reflection of Audubon's protean nature and his ability to embrace seemingly opposed stances.[39] In the *Ornithological Biography* he shows himself capable of assuming a variety of contrasting attitudes, including that of the woodsman naturalist discovering the secret haunts of birds and that of the farmer slaughtering invading rice buntings. This kind of balancing reflects Audubon's tendency to appeal to the widest possible audience and also his recognition that the only overarching narrative possible was the one in which civilization displaced wilderness, with startling speed in the case of the Ohio valley. His refusal to take a clear position in this instance ("Whether these changes are for the better or for the worse, I shall not pretend to say") suggests acceptance of a direction he viewed as inevitable, even if he was more aware than most of its costs (1:32). His ultimate response to the transformation of a country whose natural wonders he clearly loved was to appeal to the presumed immortality of art, calling on "our Irvings and our Coopers" to render the "original state" of a region forced to change so rapidly under the pressure of increasing population (1:32).

In his 1826 journal Audubon, anticipating a meeting with his idol Sir Walter Scott at the Royal Academy in Edinburgh, describes his imaginary appeals to Scott during his earlier wanderings in the woods of America as the only one who could record its primitive grandeur:

> Oh Walter Scott, where art thou? Wilt thou not come to my country? Wrestle with mankind and stop their increasing ravages on Nature, and describe her now for the sake of future ages. Neither this little stream, this swamp, this grand sheet of flowing water, nor these mountains will be seen in a century hence as I see them now. . . . Without thee, Walter Scott, she must die, unknown to the world.[40]

The prospect of meeting Scott intensified Audubon's already strong instinct for melodrama. He seems less concerned here with stopping the "ravages," although he gestures in this direction, than with preserving the image of a pristine America for the future and thus possessing through art what he could not in any other way. Ironically, his own paintings and descriptions do more to render the natural state of the country in the early nineteenth century than any account of Scott's could have hoped to, but Audubon clearly felt the need for a grand romantic narrative beyond his own powers. His subsequent appeal to "our Irvings and our Coopers" reflects a similar sense that only the best writers, in this case the leading American ones, could capture the drama and significance of the transformations he was observing. All these appeals reveal a tendency to romantic melancholy, but Audubon even in "The Ohio" looks to the expansionist future as well as to the past. He concludes this episode with the hope that the writers he invokes will record "the progress of civilization in our western country" and the deeds of the Clarks and the Boones and "many other men of great and daring enterprise" (OB, 1:32). For him belief in this progress, accompanied by heroic deeds, could coexist with grieving for the vanishing wilderness.

Modern readers may be tempted to regard Audubon as a protoenvironmentalist lamenting the transformation of the wild America he knew in many of its aspects and displaying increasing anxiety and anger at the scale of the devastation he observed in his travels in

Labrador and up the Missouri. Yet, as I have argued, Audubon was implicated to varying degrees in the practices he saw altering the environment so drastically and capable of sympathizing with many of those responsible for them. Audubon's innate theatricality may help to explain how he could assume so many contrasting attitudes: guilt at having to shoot a mother bird and pride in his ability to shoot puffins on the wing, delight at discovering the secret haunts of birds and sympathy with the farmer and the sportsman intent on killing as many as they can, melancholy over the loss of a paradisal nature and boosterish endorsement of the progress of civilization. He was acutely aware of the need to attract and entertain a diverse audience to sustain his own publishing enterprise, as the range and the exaggerations of his episodes illustrate. Perhaps many in this audience shared Audubon's ability to embrace attitudes toward the natural world that have come to seem contradictory. Perhaps they also shared the belief in the permanence of art that motivated Audubon to transform his experience into the heightened, stylized version of what he observed that one finds in his dazzling visual images. His birds offer a vision of a perfected nature, caught at a moment of dramatic intensity. Such art, like that Audubon found in a Scott or a Cooper, could make it easier to accept the inexorable receding of a wilderness that, in Pamela Alexander's words, "moves away from us steadily."

2 Henry David Thoreau and Wildness

Is it not a maimed and imperfect nature that I am conversant with? . . . Primitive Nature is the most interesting to me.

—HENRY DAVID THOREAU, *Journal,* 23 MARCH 1856

Give me a wildness whose glance no civilization can endure.

—HENRY DAVID THOREAU, "WALKING"

Henry David Thoreau began his first published essay, "Natural History of Massachusetts," by invoking Audubon:

> I read in Audubon with a thrill of delight, when the snow covers the ground, of the magnolia, and the Florida keys, and their warm sea-breezes; of the fence-rail, and the cotton-tree, and the migrations of the rice-bird; of the breaking up of winter in Labrador, and the melting of the snow on the forks of the Missouri; and owe an accession of health to these reminiscences of luxuriant nature.[1]

Audubon served Thoreau as a better illustration of the tonic effect of reading natural history than the bland surveys of the flora and fauna of Massachusetts that Emerson had asked him to review for *The Dial* and also provided a convenient springboard for his argument that one should look for health in nature rather than in society. He discovered evidence of this natural vitality in the reports from distant wild places that he found in the volumes of Audubon's *Ornithological Biography*—from the Carolinas and the great pine forest of Pennsylvania, as well as the more remote reaches of Labrador and the Florida Keys, the northern and southern limits of Audubon's ramblings. Thoreau's wide reading in natural history and the literature of exploration furnished him with abundant images of wild nature. When he imagines himself "a sojourner in wilderness," Thoreau thinks of the wilderness of northern adventurers, "by Baffin's Bay or Mackenzie's River." Yet, interestingly, he can imagine himself there only because he knows that he would find the familiar catkins of the willow and alder.[2] The most compelling evidence of the health of nature Thoreau offers in "Natural History of Massachusetts" comes from his own walks in the vicinity of Concord: the "wholesome" scent of pines; the "penetrating and restorative" fragrance of "life everlasting" in the pastures; the "strong scent of musk" blowing from the flooded spring meadows, which suggests to him an "unexplored wildness" in the landscapes of home.[3]

These landscapes, as ecologist David R. Foster has shown, were thoroughly marked by the activity of European settlers by the time Thoreau wrote, forming a patchwork of meadows, pasture, tilled fields, and woodlots.[4] The countryside he knew was much less forested than it is today. Virtually all of the woods Thoreau explored in his walks had been cut over at some point, and he saw this cutting intensify over the course of his lifetime. Thoreau's attraction to wildness and his ability to find it in familiar landscapes should be understood in the context of his exposure to continuing and extensive human use of these landscapes. He was deeply interested in the cultural patterns that had transformed the countryside and were continuing to change it and, increasingly, in the interplay between human activity and natural processes. His observations about the way pines colonized old fields and then gave way to oaks became the foundation of his argument in "The Succession of Forest Trees." Whereas Foster emphasizes what he sees as Thoreau's ability "to appreciate and even thrive on the merging of natural and cultural influences," however, I am more interested in Thoreau's persistent fascination with what he perceived as wild and in his tendency to associate the values that matter most to him with the experience of wildness.[5] Thoreau's assertion that "Primitive Nature is the most interesting to me" is part of a lament that in seeking to know all the natural phenomena that make up spring he had only an "imperfect copy," because his ancestors had "torn out many of the first leaves and grandest passages, and mutilated it in many places."[6] The sense of loss that informs this passage is a recurrent note in Thoreau's writing. While he was remarkably sensitive to the mingling of nature and culture in the landscapes around him, his perception of the impact of culture prompted efforts to recover and champion an awareness of a more primitive nature.

One of the most significant of Thoreau's many contributions to the tradition of nature writing in America was his exploration of the meaning of wilderness and wildness in America, which opened up and complicated terms that had remained largely unexamined. Thoreau's habit of thinking of wilderness and wildness metaphorically appears early, for example, in a journal entry for March 1840, in which he observes that the birds he heard singing "had for their background an untrodden wilderness—stretching through many a Carolina and Mexico of the soul."[7] It was less important to Thoreau to identify the birds ("I learned today that my ornithology had done me no service") than to imagine them as belonging to the domain of wilderness, understood as including such exotic terrain as Carolina and Mexico, and to understand wilderness as analogous to unexplored inner space. A journal entry from August 1846 demonstrates the subjectivity of Thoreau's experience of wildness as dramatically as anything in his writing. It describes a visit to Gowing's Swamp that gives rise to an extended meditation on wildness, beginning with the perception that he feels as though he has discovered a new world, "as wild and primitive and unfrequented as a square rod in Labrador, as unaltered by man," within twenty miles of Boston. By a characteristic modulation, Thoreau slides from local wildness to the wildness within: "It is in vain to dream of a wildness distant from ourselves. There is none such. It is the bog in our brain and bowels, the primitive vigor of Nature in us, that inspires that dream. I shall never find in the wilds of Labrador any greater wildness than in some recess in Concord, i.e., than I import into it."[8] Thoreau's ultimate interest was in "the primitive vigor of Nature in us," seen not only as conditioning his perception but also as breeding a consciousness that could question the habits and values associated with village life and the larger civilizing process of which it was a manifestation.

Thoreau's complex relationship to the natural world has been a central concern of critics, some of whom have examined his shifting stances toward nature and the inherent tensions between his pursuit of a higher consciousness and his desire for an intimate understanding of natural processes.[9] Thoreau's preoccupation with wildness has become a familiar topic of critical discourse, yet more can be said about the evolution of his sense of wildness and its implications and about how this was shaped by his ongoing encounters with the natural world.[10] Thoreau's understanding of "the primitive vigor of Nature in us" was nourished by his perception of what he took to be wildness in nature as he continued to explore his immediate environment. While an interest in wildness can be found throughout his writing, from early journal entries to his late essays, this interest peaked in writings from the mid-1840s through the mid-1850s. My primary focus will be on the published works in which Thoreau explores wilderness and wildness in the most sustained way, *The Maine Woods* (1864) and "Walking," and on journal entries that reveal patterns in his thinking about wildness, although I consider writing from throughout his career, including his late essays.

One can get a sense of what Thoreau saw as manifestations of wildness in the natural environment that he explored in his walks around Concord and of how strongly he responded to these from his early journals and the essays that drew upon them. The pine tree emerges early as an emblem of wildness: "The pine stands in the woods like an Indian—untamed—with a fantastic wildness about it even in the clearings."[11] Here Thoreau endows the pine with the independence and the primitive character he associated with the Indians of his imagination ("The pitch pines are the ghosts of Philip and Massasoit"). To the eye adapted to village trees such as maples and oaks, its shape appears "fantastic" and resistant to domestication. Thoreau would come to identify with the pine himself ("That tree seems the emblem of my life—it stands for the west—the wild") and would make it a symbol of the wildness of the Maine woods, as well as of the wilderness the farmer must subdue ("The civilized man regards the pine tree as his enemy").[12] In another early journal entry, he writes that "wildness as of the jay and the muskrat reign[s] over the great part of nature."[13] Muskrats make frequent appearances in the journals, sometimes as a symbol of wild nature coexisting with civilization. The raucous cry of the jay also recurs in the journals as a symbol of wildness.[14] Thoreau would associate the song of other birds with wildness as well, particularly that of the wood thrush, which offers a different kind of strangeness, beguiling rather than harsh: "There is a sweet wild world that lies along the strain of the wood thrush."[15]

Thoreau's attraction to wild nature appears to be governed by aesthetic ideals at times, particularly in his early writing, as in his appreciation of the mellifluous song of the wood thrush, his acute pleasure in the fragrance of the sassafras, and his delight in the look of the woods ("There is something indescribably wild and beautiful in the aspect of the forest skirting and occasionally jutting into the midst of new towns").[16] As this last example suggests, the aesthetic appeal of the wild often depends upon its contrast with the domestic and familiar. In a journal entry from April 1852, Thoreau comments on the way a fence enhances the appeal of an oak and the undulating half-acre in which it stands. The effect would be lost in an unfenced prairie, he notes, but the fence "frames it & presents it as a picture."[17] Here the actual framing of a scene makes it picturesque for Thoreau. Elsewhere, picturesque effects are created by the mingling of wild and domestic, as in a view of woodcutters with their oxen on the ice of Walden Pond that Thoreau describes in a journal entry from the same year as "a pretty forest scene . . . a piece of still life."[18]

Thoreau plays on this contrast between the wild and the domestic in *A Week on the Concord and Merrimack Rivers* (1849), where these elements often mingle, although he shows more interest in this work in the benefits to be gained from exposure to wildness. When he incorporated the sentence about the forest jutting into the town into *A Week,* he changed "wild and beautiful" to "inspiriting and beautiful," accenting what he saw as the beneficial influence of forest on the town. He would develop this theme more fully in "Walking." In *A Week* he associates the straightness of the trees with the "ancient rectitude and vigor of nature" and asserts that we need "the relief of such a background, where the pine tree flourishes and the jay still screams."[19] Thoreau at least needed such relief, proof for him of the inescapable pressure of nature; he wanted the forest to penetrate the boundaries of the town and the jay's screams to intrude on the consciousness of its inhabitants.

Wild nature was of course more than background for Thoreau. In his journal for January 1844, he describes himself skating after a fox that appeals to him for belonging to "a different order of things than that which reigns in the village."[20] Thoreau's early writing reveals his attraction to this "different order of things" and also the sense in which perceiving wildness depends upon recognizing and valuing difference. In the *Journal* one can see Thoreau learning to turn this sense of difference against conventional modes of living and thinking. His perception of the free and elusive ways of the fox triggers a critique of the village mindset in which the fox becomes vermin, to be controlled by bounties, and is reduced to an example of "proverbial cunning." By observing the fox's actual behavior, praising its wildness, and engaging it in a form of play Thoreau distinguishes himself from the village world while at the same time recognizing that he cannot truly enter the world of the fox.

The instinctual behavior of the fox illustrates one kind of freedom that Thoreau associated with wildness. Another can be seen in the unchecked growth of areas that have not come under human control. Although Thoreau recorded the local rituals by which nature was appropriated for human use, such as ditching wet ground and mowing the meadows, his imagination was most engaged by examples of undomesticated nature. An encounter with his "shiftless" neighbor John Field, going slowly about his work of "bogging" meadows, prompts a wish to see meadows left wild and an imaginative leap to the freedom of a mind unconstrained by orthodox education, as if he is intrigued by the possibility of a primitive spontaneity of thought that resists domestication: "Grow wild and rank like these ferns and brakes, which study not morals nor philosophy, nor strive to become tame and cultivated grass for cattle to eat."[21] Thoreau's injunction points to still another characteristic he associated with wildness—rank growth reflecting the unruly fertility he found in undisturbed nature. Swamps provided the most dramatic examples of rankness for Thoreau. He can declare in "Walking" that he enters the swamp as "a sacred place, a *sanctum sanctorum*," because he finds in its biological richness "the strength, the marrow of Nature."[22] The swamp becomes a place where he can "recreate" himself because its teeming vitality suggests the possibility of a vibrant life that Thoreau understood as spiritually as well as physically invigorating, outside the dull and ordered world of the village.

Praising the "dismal" swamp, even suggesting that it be substituted for the front yard, suited Thoreau's rhetorical strategy in "Walking" of taking extreme positions to challenge the received opinions of the "citizen," yet swamps consistently embody the virtues of wildness for him. The frequent accounts in the *Journal* of excursions to swamps of a variety of kinds reveal the depth of Thoreau's attraction to them and some of the sources of their appeal. He seems to have needed to keep returning, as if to renew his sense of the pungency and fertility of nature in its primal form. His description of a visit to Beck Stow's swamp just outside Concord, in July of 1852, suggests how the isolation and strangeness of a place beyond the reach of cultivation make it a source of spiritual nourishment:

> When life looks sandy & barren—is reduced to its lowest terms—we have no appetite & it has no flavor— Then let me visit such a swamp as this deep & impenetrable where the earth quakes for a rod around you at every step.—with its open water where the swallows skim & twitter—its meadow & cotton grass—its dense patches of dwarf andromeda now brownish green—

with clumps of blue-berry bushes—its spruces and ver-
durous border of woods imbowering it—on every
side.[23]

An early visit to Miles's swamp prompts the observation that nature
excels art in its "luxury and superfluity" and produces its own fantastic shapes. Encountering nature's "luxurious and florid style" in
the swamp leads to the assertion that "she is mythical and mystical
always," a claim that Thoreau made part of the argument of "Natural
History of Massachusetts."[24] After a visit to the same place eleven
years later, Thoreau comments that swamps are "the wildest and
richest gardens that we have. Such a depth of verdure into which
you sink," characteristically inverting the conventional understanding of "gardens" by associating the term with pristine nature rather
than with nature refined by the gardener's art.[25]

What the gardener would regard as excessive, "rank" in the
sense of being out of control, Thoreau prized, whether he found it
in swamps or in other relatively undisturbed natural settings: the
rich mud and "luxuriance of weeds" of river banks, the "fecundity
and vigor" of masses of Joe Pye weed and goldenrod through which
he wades in the late summer.[26] He reacted to the sight of newly
mown meadows (in his journal for August 1852) by declaring: "I
would like to go into perfectly new and wild country where the
meadows are rich in decaying & rustling vegetation—present a
wilder luxuriance—I wish to lose myself amid reeds & sedges &
wild grasses that have not been touched."[27] Thoreau's fantasy of losing himself in wild grasses implicitly repudiates the gardener's aesthetic and the stance of detached admirer of an improved nature
that this implies. In the luxuriance of swamps and unmowed meadows Thoreau found a natural vitality that he would experience as
directly and fully as possible. Local landscapes provided his own
version of the "luxuriant nature" that engaged him in Audubon's
narratives of exotic places.

In a frequently quoted passage from his journal for June 1840,
Thoreau imagines himself spending the day submerged in a swamp:
"Would it not be a luxury to stand up to ones chin in some retired
swamp for a whole summer's day, scenting the sweet fern and bil-

berry blows, and lulled by the minstrelsy of gnats and mosqui-toes?"[28] Thoreau was capable of moods in which swamps could be too "dismal and dreary" even for him ("I would as lief find a few owls and mosquitoes less"), but his characteristic mood is one of absorbed delight in the strong sensations he found in them, includ-ing the "rank smells" of fecundity.[29] Immersion in a swamp, or wad-ing in a marsh ("We need the tonic of wildness,—to wade some-times in marshes where the bittern and the meadow-hen lurk, and hear the booming of the snipe"), becomes a figure for the kind of intimacy with wild nature that Thoreau craved, one that we see him seeking in various ways in the experiences that he describes in his journals.[30]

The adjective "primitive" functions as another snynonym for "wild" in Thoreau's vocabulary, along with "free" and "rank." A weedy river bank suggests the "South American primitive forests" of his reading and, with its "antediluvian rocks," seems a fit habitat for "primitive wading birds."[31] The word resonated for Thoreau because his imagination was drawn so strongly to the natural world that existed in America prior to European settlement. His desire to experience this world frequently surfaces in his writing, as in his description of a brook he finds flowing in "a recess apparently never frequented," just as it did a thousand years ago: "It is a few rods of primitive wood—such as the bear & deer beheld—It has a singular charm for me in carrying me back in imagination to those days."[32] Thoreau reports discovering a fisherman's box in the brook, but only after he has indulged his fantasy of the place as unfrequented before the moment of his arrival. The hooting of the owl appeals to him as something his "red predecessors" heard in the same place more than a thousand years before, a "grand, primeval, aboriginal sound."[33]

Thoreau's lifelong fascination with "the Indian" is one measure of his desire to recover a sense of the wildness of primitive Amer-ica. His efforts to imagine presettlement conditions in *A Week* reflect a conventional and idealized understanding of savage life that his experiences in Maine with his Penobscot guides Joe Aitteon and Joe Polis, with their adaptations to white ways, would greatly compli-cate.[34] By giving primacy to the original name of the Concord

River (the Musketaquid or Grass-ground) at the outset of *A Week,* Thoreau establishes his sympathy for a more primitive and natural understanding of place: "It will be Grass-ground River as long as grass grows and water runs here; it will be Concord River only while men lead peaceable lives on its banks."[35] What the early settlers saw as "howling wilderness" and transformed to orchards and gardens, the Indians regarded as home, he observes of the environs of Billerica, where he and his brother spent their first night on the river. Thoreau attributes to the Indian a kind of life he clearly envies: "By the wary independence and aloofness of his forest life he preserves his intercourse with his native gods, and is admitted from time to time to a rare and peculiar society with Nature."[36] What he describes as "a singular yearning toward wildness" in his own nature seems to be stimulated by the evidence of civilization, in this case the village spire and fields on either side of the river with "a soft and cultivated English aspect."[37] One cannot be understood apart from the other. Imagining the primitive became for Thoreau a means of critiquing what he saw as overly refined and ordered and, increasingly, of exploring unmapped regions in his own consciousness.

Thoreau's "yearning toward wildness" found expression in gestures toward a more primitive life, beyond the famous one of building a hut and living for a little more than two years beside Walden Pond, which his neighbor Hawthorne characterized as leading "a sort of Indian life among civilized men."[38] Thoreau was attracted to the idea that we have a savage within us and perhaps an "original wild name" in addition to our familiar one, like the "savage who retains in secret his own wild title earned in the woods."[39] His sense of primitive life extended beyond the local example of the American Indian to instances he encountered in his reading, such as Tahitians and Africans, and especially to the heroes of Ossian's supposedly ancient poetry. The latter impressed Thoreau by their "simple, fibrous life," which illuminated by contrast the debility and luxury of the life chronicled by "our civilized history." He sought to validate his own efforts to recover the primitive by insisting that our connection with that life remained vital: "Inside the civilized man stands the savage still in the place of honor. We are those blue-eyed, yellow-haired Saxons, those slender, dark-haired Normans."[40]

What Thoreau found in such examples of primitive life, ignoring cultural differences, was a kind of health and vigor that he associated with nature itself. His efforts to undo the effects of civilization in his own life took the form of sometimes startling gestures toward the savage, none more startling than the one with which he opens the "Higher Laws" chapter of *Walden:* "I caught a wild glimpse of a woodchuck stealing across my path, and felt a strange thrill of savage delight, and was strongly tempted to seize and devour him raw; not that I was hungry then except for that wildness which he represented" (1854).[41] The example offers a way for Thoreau to recognize the tension in himself between an instinct toward a higher, spiritual life and that toward a "more primitive rank and savage one" and the larger tension between sensuality and purity that he goes on to explore in the chapter. In developing an ideal of purity that does not allow for "unclean" animal food and linking this with his own evolution from hunter and fisherman to poet and naturalist, Thoreau may seem to undermine the gesture with which he begins the chapter. Yet for all the austerity that he urges in one's personal life, with its emphasis upon strictly disciplining sensual appetites, Thoreau recognizes that our natures are a mixture of the divine and the animal.[42] He can repudiate "unclean" personal habits and still legitimize attraction to a wildness that causes him to want "to take rank hold on life and spend my day more as the animals do." His rhetorical strategy is to insist, at the outset, that desires we might regard as contradictory are in fact compatible: "I love the wild not less than the good."[43]

The *idea* of eating a woodchuck raw was sufficient for Thoreau, although he did try eating one cooked, describing this experiment in an earlier chapter of *Walden.*[44] He also sampled boiled acorns. The wild foods that Thoreau actually ate, especially the berries and wild apples that he regularly sought out, fed the imagination as well as the body with tastes that could be sharp and strange, like that of the teas made from available plants that he describes himself drinking in *The Maine Woods.*[45] Thoreau craved what he called in his late essay "Wild Apples" the "sours and sweets of nature," properly enjoyed only in the field, and took pride in his appetite for them: "It takes a savage or wild taste to appreciate a wild fruit."[46] He delighted in wild apples so seasoned by the weather that "they *pierce* and *sting* and *permeate* us with their spirit."[47] Thoreau clearly wanted an intensity that he could not find in the "tame and forgettable" taste of domesticated apples. The piercing taste of wild ones became a metaphor for an awakening experience.

Responsiveness to wild tastes as well as to the varied odors he encountered in the natural environment, to the rank smell of the swamp along with the fragrance of the sweet fern, was a way for Thoreau to experience something of the wildness that attracted him. He associated himself with pine trees and lichens ("Our spirits revive like lichens in the storm") and imagined himself happily submerged in a swamp all day, seeming to want to identify completely with wild nature.[48] Yet he could also recognize the differences that separated him from the natural world. He inevitably loses the game of trying to get closer to the loon on Walden Pond whose "wild laugh" fascinates and seems to taunt him.[49] Thoreau describes himself acutely in "Wild Apples" as being like the wild apple in not belonging to "the aboriginal race" but "having strayed into the woods from the cultivated stock," acknowledging his links to the culture from which he tries in various ways to distance himself.[50] The combination of a powerful attraction to wildness and a consciousness of the barriers to attaining it generates much of the tension in his prose.

Thoreau's renderings of his three extended trips to Maine (in 1846, 1853, and 1857) in *The Maine Woods* reveal both his fascination with wilderness of a kind that he could not experience in Massachusetts or New Hampshire and the extent to which his "cultivated" origins and his continuing investment in the civilized life of towns conditioned his perceptions. The Maine journeys tested Thoreau's capacity to respond to wilderness on a vast scale like nothing else in his experience except perhaps his encounters with the "vast and wild" ocean on his trips to Cape Cod. The sea was for him another kind of "wilderness," "wilder than a Bengal jungle, and fuller of monsters."[51] With its violence, evidenced in the shipwreck Thoreau graphically describes in the opening chapter of *Cape Cod* (1865), the sea revealed another dimension of wild nature, alarming in its destructiveness but fascinating nonetheless.

In "Ktaadn," the first of the Maine essays, one finds a mixture of delight and unease in Thoreau's descriptions of his encounters with wilderness. Proceeding inland by stagecoach from Bangor, he responds to the "beauty" of the varied evergreens lining the road but sees beyond this border "the grim, untrodden wilderness, whose tangled labyrinth of living, fallen, and decaying trees only the deer and moose, the bear and wolf can easily penetrate."[52] Adjectives such as "grim" and "untrodden" say more about Thoreau's expectations than about what he saw. He does convey a strong sense of being enclosed by a dense forest inhospitable to human presence, for example, when he imagines a solitary sled track running up into the winter wilderness "hemmed in closely for a hundred miles by the forest" (*MW*, 33). Clearings, some extensive, demonstrate the progress of settlement, but Thoreau is always conscious of the immense, uninterrupted forest beyond. The minimal clearings of the loggers' camps, surrounded by "drear and savage" scenery, suggest how provisional the human presence could seem in this environment (*MW*, 19–20). In the conclusion of the essay Thoreau celebrates the wild, "unsettled and unexplored" character of much of Maine, picturing it as representative of American wilderness, "where the Indian still hunts and the moose runs wild" (*MW*, 82) and "there still waves the virgin forest of the New World" (*MW*, 83). Yet he continues to call this country "grim" and "universally stern and savage" in aspect, even invoking the cliche "howling wilderness" (*MW*, 82–83). Thoreau ends the essay by attempting to strike a balance between a view of wilderness as alien and threatening, responding to the conventional expectations of his audience as well as registering his own unease, and praise of flora and fauna that he found immensely appealing.

Thoreau's closing generalizations about the "inexpressible tenderness" and "perpetual youth" of a "blissful, innocent Nature" (*MW*, 81) seem forced. When he looks closely at these woods, however, his language becomes more convincing. In Maine Thoreau found an intensification of kinds of wildness that appealed to him in familiar landscapes, including "bracing" fragrances and, especially, a pervasive wetness: "The primitive wood is always and everywhere damp and mossy, so that I travelled constantly with the impression that I was in a swamp" (*MW*, 20). Moisture was a fundamental attribute of wildness for Thoreau, as it was for Gerard Manley Hopkins ("What would the world be, once bereft / Of wet and of wildness?").[53] In *A Week* Thoreau characterizes the Pemigewasset, which rises in the White Mountains of New Hampshire, as flowing "through moist primitive woods whose juices it receives" and imagines himself being "soaked in the juices of a swamp."[54] The "forest houses" of the loggers in Maine fascinate him for the way they blend with their setting, with their green logs "hanging with moss and lichen . . . and dripping with resin, fresh and moist, and redolent of swampy odors, with that sort of vigor and perennialness even about them that toadstools suggest" (*MW*, 20). The houses embody the quintessential juiciness of the woods and a vigor and promise of renewal that Thoreau associated with moisture, whether in woods or swamps or wet meadows. In "Chesuncook," his second Maine essay, he explores the way human occupation changes the character of a forest, destroying the "wild, damp, shaggy look" he admired in the Maine woods and leaving swamps as the only remaining "primitive places" (*MW*, 150). Clearing the land meant drying it out and thus eliminating the rank vitality that was the source of its appeal.

Thoreau's description of the beer his party was offered at Thomas Fowler's house, "strong and stringent as the cedar-sap," registers his desire to absorb the kind of wildness he was discovering in the Maine woods. When they start on their expedition, poling up the rapids of the Penobscot proves "exhilarating." Yet the excitement that stimulates Thoreau's imagination has an edgy quality in his account of rowing across North Twin Lake in the moonlight. He exults in the absence of other humans ("No face welcomed us but the fine fantastic sprays of free and happy evergreen trees" [*MW*, 36]) but goes on to comment on the difficulty of finding the outlet at the far end of the lake and to report an anecdote of experienced woodsmen "lost in the wilderness of lakes" (*MW*, 37). His description of how some "utterly uncivilized" owl "hooted loud and dismally in the drear and boughy wilderness, plainly not nervous about his solitary life" (*MW*, 38) suggests that humans in such a "wilderness" have something to be nervous about. Singing their Canadian boatsong ("The Rapids are near and the daylight's past!") with "new emphasis" becomes a way that he and his companions make a space for themselves in the wilderness night. Thoreau describes himself as waking at midnight in the campsite they eventually reach, their first, and thinking that "the little rill tinkled the louder, and peopled all the wilderness for me" (*MW*, 40), as if conscious of a need for a familiar, reassuring sound to fill the engulfing silence. His observation that the scene of "dark, fantastic rocks" rising from the calm surface of the lake left a strong impression of a "stern, yet gentle wildness" captures the mingled feelings that the experience arouses. The adjective "fantastic" usually has positive connotations for Thoreau, suggesting a welcome deviation from aesthetic norms. Here, coupled with "dark," it suggests the potential of the scene to become menacing. If Thoreau were truly solitary, it might have. Unlike Audubon in his early ramblings and John Muir in his Sierra explorations, however, Thoreau did his wilderness adventuring in the company of friends and guides. It was a gentlemanly kind of travel, an early form of ecotourism that he describes with a travel writer's eye for detail for readers who might follow him or simply enjoy the experience vicariously.

One of the striking things about Thoreau's rendering of wilderness scenes in "Ktaadn" is his tendency to imagine them in relation to an advancing civilization. Upon entering North Twin Lake at sunset, he perceives it as lying "open to the light with even a civilized aspect, as if expecting trade and commerce, and towns and villas" (*MW*, 36). Thoreau found lake views "mild and civilizing" (*MW*, 80), providing a sense of relief from the oppressiveness of the "uninterrupted" forest.[55] Perhaps he also regarded the appearance of lakes as "civilizing" because they conformed to notions of the picturesque, as dense woods could not, but it is interesting that he instinctively imagines future human settlements ("towns and villas"). He seems to have been able to celebrate the wild character of the country more readily because he could understand it in the context of an expanding and encroaching civilization. While he found the evidence of loggers his party encountered "startling," such signs of human presence (a ring bolt, a barrel of pork) enabled him to imagine interactions with wilderness that made it seem less overwhelming. Coming down Katahdin after his abandoned effort to reach the summit, he speculates on how the cranberries his companions are gathering may be the basis for commerce, "when the country is settled and roads are made," and responds to the panoramic view by imagining the land below as "a large farm for somebody, when cleared" (*MW*, 66). The assumption that settlement will come is not surprising, given the rapidity with which it was spreading at the time, but in this context such reflections suggest an effort of the imagination to recover from the disconcerting experience he describes among the clouds and rocks above.

Thoreau's rendering of his attempted ascent of Katahdin has elicited a broad spectrum of critical readings, with one extreme marked by arguments that the experience was traumatizing, or at least represented a defeat of the imagination, and the other by the contention that it should be seen as a revelatory encounter with the sublime.[56] It is clear that Thoreau shaped his account of the episode, retrospectively, to maximize its drama.[57] The experience that he gives such dramatic power is atypical of his encounters with wild nature, even of his experience of mountains. In *A Week* Thoreau gives relatively little attention to his ascent of Mt. Washington (Agiocochook) and makes no comment on his experience of its bleak summit. His account of a subsequent ascent of Saddleback Mountain in 1844, on the other hand, emphasizes a dawn vista of undulating clouds that becomes "such a country as we might see in dreams, with all the delights of paradise." The scene stimulates Thoreau to imagine a transformed earth, "with no spot or stain."[58] One should not exaggerate the importance of the Katahdin experience to Thoreau's career as a writer, given his continuing interest in mountains and his varied reports of summit experiences. It is significant, nevertheless, for what it reveals about the boundaries of his empathy for wild nature in its starkest form.

Thoreau's account of his initial foray toward the summit of Katahdin, before retiring to spend an unsettled night with his companions, shows a kind of assurance missing in the account of his ascent the next morning. Alluding to Satan's journey through Chaos in *Paradise Lost* in describing his difficult progress over the tops of stunted spruces allows him to give his efforts something of the mock-heroic character of Milton's account of Satan's adventuring ("With head, hands, wings, or feet . . . / And swims, or sinks, or wades, or creeps, or flies").[59] Thoreau presents himself, unflatteringly, as having "slumped, scrambled, rolled, bounced, and walked, by turns, over this scraggy country," then ironically shows the disjunction between familiar landscapes and the boulder-strewn terrain he finds by summoning a mock-pastoral vision of "flocks and herds . . . chewing a rocky cud at sunset": "They looked at me with hard gray eyes, without a bleat or low" (*MW,* 61). This capacity for irony disappears in his account of his experience of fog and rocks on the "table-land" just below the summit the next morning. Here is Chaos indeed, a second allusion to *Paradise Lost* suggests, disorienting and seemingly hostile to any kind of human presence. No satisfying imaginative transformation of this "raw," "unfinished" landscape seems possible.

Thoreau represents his solitariness as weakness, not as a precondition for enlightenment as it might be in other circumstances: "Vast, Titanic, inhuman Nature has got him at disadvantage, caught him alone" (*MW,* 64). Milton's Satan smoothly talks his way past the challenges of "Chaos and ancient Night," presenting himself as a solitary adventurer in need of help ("Alone, and without guide, I seek / What readiest path leads where your gloomy bounds / Confine with Heav'n").[60] Thoreau, by contrast, has no answer to Nature's imagined questions and accusations, portraying himself as an intruder and turning back as if from forbidden ground. This is a failed quest, unlike Satan's. The experience Thoreau describes here recalls Audubon's account of feeling increasingly alienated from the "barren lands" of Labrador, as he becomes more conscious of a "deep silence" and the lack of human presence. Thoreau probes his experience more deeply, of course, but the encounters resemble each other in suggesting how wildness can become desolation when a landscape seems to offer little or no basis for making imaginative connections with familiar human experience.

Reacting to another kind of desolation in "Burnt Lands," Thoreau comments on the difficulty of imagining "a region uninhabited by man." He can define such a place only by negations, opposing it to the humanized landscapes of his former experience: "Here was no man's garden, but the unhandselled globe. It was not lawn, nor pasture, nor mead, nor woodland, nor lea, nor arable, nor waste-land" (*MW,* 70). If this is an example of the romantic sublime ("Nature here was something savage and awful, though beautiful"), Thoreau seems disinclined to celebrate it. Such "primeval, untamed, and forever untameable *Nature*" (*MW,* 69) is unsettling because he cannot imagine it coexisting with or preparing the way for civilization. It is not "wild" in any of the senses in which Thoreau usually understood the word and thus not containable by his customary opposition of wild and domestic. The sense of crisis arises from Thoreau's experience of encountering what he understands as raw matter ("vast, terrific") and finding his conception of himself shaken in the process: "*Contact! Contact! Who* are we? *where* are we?" (*MW,* 71) The sense of separation between the conscious mind and the body that Thoreau represents himself as experiencing ("I fear bodies") undermines the delight in the sensuous apprehension of physical nature and the ability to find spiritual meaning in it that normally drive his celebrations of wildness.

Thoreau's reflections upon the conclusion of his expedition back at Tom Fowler's show him recovering a positive sense of primitivism in his romantic evocation of the settler living "on the edge of the wilderness," playing his flute in the evening to the accompaniment of wolf howls, and the even more romantic vision of the "ancient and primitive Indian" that he imagines as a "dim and misty" figure gliding up the Millinocket in his bark canoe, more acceptable than the fact of Louis Neptune finally ready to act as guide after emerging from his "drunken frolic" (*MW,* 78). The relief at returning to the mixed landscape around Concord that Thoreau expresses at the end of "Chesuncook," based on a trip taken seven years later, shows a settling of his understanding of the Maine wilderness and its place in his experience. He now confidently identifies with the poet who thrives in a "partially cultivated country" of woods and fields, "with the primitive swamps scattered here and there in their midst" (*MW,* 155). This is a "humanized" nature, made possible by the logger and pioneer, with remnants of wilderness such as pines and orchids to remind the poet of the need for periodic renewal "at some new and more bracing fountain of the Muses, far in the recesses of the wilderness" (*MW,* 156).

In "Chesuncook" and also in "The Allegash and the East Branch" (based on his third trip to Maine in 1857) Thoreau focuses more closely on botanical detail, as in the later journals, and shows himself more comfortable with the Maine wilderness, though not at home in it. Upon arriving at the carry from Moosehead Lake to the Penobscot River to begin the excursion described in "Chesuncook," he delights in the "wild and primitive look" of the northern flora and finds the evergreens that crowd the sides of the carry welcoming (*MW,* 93). The phosphorescent wood that he describes observing in the fire in "The Allegash and East Branch" seems "a light shining in the wilderness for me" (*MW,* 180), and he describes lying awake on the shore of Chamberlain Lake listening to the "thrilling" cry of the loon "give voice" to the wildness of the place (*MW,* 224). This after he and his companion have spent much of the day trying to find their way over Mud Pond Carry, some of it slogging through a tangled cedar swamp harassed by black flies. The ironic frame that Thoreau gives that experience in retrospect, evident in his observation that it is the imagination of the traveler that howls and not the "howling wilderness" (*MW,* 219), reveals a capacity for detachment and self-criticism not apparent in "Ktaadn."

In the later Maine essays one sees Thoreau absorbed in learning from the botanical richness of the woods and the ways of his Penobscot guides. Lying in camp with Indian moose hunters, listening to them speak Abenaki, he takes pleasure in hearing the "purely wild and primitive American sound" (*MW,* 136) even though he understands nothing. In Joe Polis, his guide for the last of his Maine trips, Thoreau found a particularly skilled and engaging tutor. His attitude toward Polis, whom he persists in calling "the Indian," is close to reverential: "Nature must have made a thousand revelations to them which are still secrets to us" (*MW,* 181). Polis's "wilderness walk" of three days to Oldtown impresses him as "traveling of the old heroic kind over the unaltered face of nature" (*MW,* 235). Thoreau proves an eager pupil, fascinated with Polis's woodcraft and his use of Indian words, but not always an adept one. When Polis returns to rescue him and his companion from their unheroic adventure at Mud Pond Carry, he expresses surprise that they were unable to follow his tracks. With his minimalist approach to living in the woods—with no change of clothes and only an axe, a gun, and a blanket—Polis offers a revealing contrast to the heavily laden Thoreau. Despite his enthusiasm and his growing familiarity with backcountry Maine Thoreau continues to be marked as a tourist, in ways that Muir in his Sierra rambles never is.

The other side of Thoreau's fascination with evidence of wildness (and primitivism) in the later essays is a sharper sense of the destructiveness of human uses of wilderness, apparent in his revulsion at the waste of moose hunting, focused for him by the "tragical business" of skinning a moose (*MW,* 115), and in his denunciations of logging. Not surprisingly, attitudes toward the pine tree become the test of the right use of wilderness, and "the poet" sets the standard: "It is the living spirit of the tree, not its spirit of turpentine, with which I sympathize, which heals my cuts" (*MW,* 122). The logger becomes the agent of death and commerce, dealing in carcasses of pines and in pine-tree shillings, and the vast, continuous forest of the end of "Ktaadn" has metamorphosed into a wilderness under attack by "10000 vermin gnawing at the base of her noblest trees" (*MW,* 228).

Thoreau was capable of being intrigued by the ways of loggers in the woods, around Walden Pond as well as in Maine, but the vision of a war against the pines that he offers in his last two Maine essays is consistent with the defense of wildness that he develops in the course of his writing and gives strongest expression in "Walking." This vision helps to explain the call for national wilderness preserves, for "inspiration" and "true recreation," with which Thoreau concludes "Chesuncook." His proposal, and his rudimentary effort to establish a rationale for such preserves, marks the real beginning of the argument for wilderness preservation in America.[61] This preservationist impulse distinguishes Thoreau from Audubon and others who lamented the rapid disappearance of wilderness but fatalistically accepted this as an inevitable cost of progress.

Wilderness of the sort he discovered in Maine remained an ideal for Thoreau, haunting his imagination. Seeing Fair Haven Pond in the moonlight could dissolve the reality of nearby village and farms and give at least the illusion of inhabiting a primitive America: "Fair Haven by moonlight lies there like a lake in the Maine Wilderness in the midst of a primitive forest untrodden by man. This light & this hour takes the civilization all out of the landscape—."[62] Thoreau found various ways of taking civilization out of the landscape by an act of imagination, as one sees in his misty vision of the Indian gliding up the Millinocket in his bark canoe or in his descriptions in the journals of scenes in which fog masks the features of the settled landscape. One of the appeals of night walks for Thoreau was the way they diminished his sense of human intrusion on the landscape, freeing the imagination to dwell on a wildness prior to civilization. On one such walk he finds "something creative & primal in the dewy mist," suggesting to him the fertility he associates with the origin of things, "where the ancient principle of moisture prevails."[63]

Thoreau's powerful argument for the "tonic" of wildness in the "Spring" chapter of *Walden,* little changed from the initial draft of 1846–47, mingles evocations of wilderness with impressions from wading in familiar marshes: the "booming of the snipe," the smell of "the whispering sedge where only some wilder and more solitary fowl builds her nest."[64] Thoreau presents such intimate experiences of wildness as an antidote to the stagnation of village life, yet he insists on the need to preserve a sense of the mysteriousness of a nature "infinitely wild, unsurveyed and unfathomed by us because unfathomable," and associates this with the "inexhaustible vigor" and "vast and Titanic features" of nature on the scale of the ocean and "the wilderness with its living and its decaying trees" (*Walden,* 318). Thoreau had sufficiently assimilated his sometimes unsettling experience of wilderness on his first Maine expedition and his encounters with an immense and indifferent sea on Cape Cod to understand confronting nature on this scale as tonic, invigorating us and at the same time reminding us of our limitations: "We need to witness our own limits transgressed, and some life freely pasturing where we never wander" (*Walden,* 318). For Thoreau to invest the natural world with the meaning he does, he needed to endow at least some parts of it with the freedom and power to resist human intervention and the mysteriousness to defeat our attempts to measure and thus know it.

Thoreau did not need his experience of wilderness to find

assurance of "the appetite and inviolable health of Nature" in a dead horse by the path to his house at Walden or evidence of nature's "living poetry" in the thawing sand of the nearby railroad cut. In elaborating on the "sand foliage" passage in revisions of his initial draft of the "Spring" chapter, Thoreau emphasizes the way the "hieroglyphic" of the flowing sand revealed patterns to be found throughout nature (308). It also revealed the immense, inexhaustible energies of nature. The sense of the health and vigor of nature that Thoreau conveys in this chapter and elsewhere in his work is critical to his sense of the value of wildness and informs the central argument of "Walking."

Thoreau's most provocative argument for wildness and his richest exploration of its implications can be found in "Walking," not published until 1862 but substantially complete by 1857.[65] His famous declaration, "In Wildness is the preservation of the World," assumes that wild nature embodies a primitive and self-renewing fertility and the health and vigor that flow from this and that it nourishes a freedom of thought and imagination not possible for one bounded by the ordered nature and conventional habits of village life. To sustain his central argument, Thoreau must see wild nature as the measure of vitality ("The most alive is the wildest") and as capable of acting upon those who come in contact with it.[66] He assumes that "forest and wilderness" are indeed tonic, the swamp invigorating, for others as for him. And not only for individuals but for society: "A town is saved, not more by the righteous men in it than by the woods and swamps that surround it" (229). Salvation, and preservation, imply for Thoreau a reformation of consciousness stimulated by contact with a wild nature that is symbolized by the primitive forest he imagines rotting beneath Concord and sustaining it as other such forests had Greece and Rome. Economic health was not Thoreau's concern, except in the negative sense that it reflected the acquisitive instincts that he consistently satirized. His measure of culture and spiritual health was the capacity of a town, or a civilization, to produce poets and philosophers. The figure of John the Baptist emerging from the wilderness eating locusts and wild honey can serve as his emblem of "the Reformer," rather than as prophet of a redeemer to come, because of the generative power with which he invests wilderness. Thoreau locates the kind of salvation he values in wild food and the participation in wildness that the act of eating it suggests.

The notion of wildness becomes most powerful in Thoreau's writing when he exploits it metaphorically, as he does repeatedly in "Walking." This power depends upon understanding wildness partly as "the primitive vigor of Nature in us" and as the source of freedom in thought and action, whether the "uncivilized free and wild thinking" he finds in *Hamlet* and *The Iliad* or the "original wild habits and vigor" of a cow who breaks out of her pasture. His dismissal of most of English literature as "tame and civilized" reflects a preference for the "pristine vigor" of a literature more attuned to mythology (like that of ancient Greece) because more directly connected with a nourishing natural world. English literature, like English lawns, suffered from a refinement that distanced it from standards derived from the nature that he saw as the source of truth.

Thoreau repeatedly asserts the standard of primitive nature in "Walking," what he characterizes at one point as "this vast, savage, howling mother of ours, Nature, lying all around, with such beauty, and such affection for her children, as the leopard" (237), emphasizing its savagery to heighten the contrast with civilization. The ideals he advocates depend upon reversing, or thwarting, a civilizing process that he saw as producing at best "a merely English nobility," inbred and without future. Children should learn a wild and dusky knowledge, a "tawny grammar," from their leopard mother. Poets should attempt to "[nail] words to their primitive senses"; such words should be "so true and fresh and natural" that they appear to swell like buds as spring approaches (232). A capacity to reflect the vigor of nature becomes the standard for poetic language. Thoreau's statement that he knows no poetry that adequately expresses "this yearning for the Wild" reveals an awareness that his images describe something closer to a yearning than to an attainable ideal.

The boldness of Thoreau's rhetorical stance in "Walking" and of his efforts to link human understanding and imagination with wild nature metaphorically reflect his sense of the difficulty of undoing the effects of civilization. His description of wildness as "not yet subdued to man" (226) acknowledges the strength of the tide against which he has set himself. The engagingly subversive paragraph that he begins by playing on the language of the Creed ("I believe in the forest, and in the meadow, and in the night in which the corn grows") ends with a mock prayer for "a wildness whose glance no civilization can endure,—as if we lived on the marrow of koodoos devoured raw" (225). The power of nature, expressed as an infusion of arborvitae in tea or as the marrow of antelopes, is analogous here to the prophetic voice of an Isaiah. Thoreau needed to invoke this kind of power to find a voice that his civilized audience could not ignore.

Thoreau made his "extreme statement" in "Walking" and set himself against the "champions of civilization" by exaggerating the opposition of natural and civil. To declare man "a part and parcel of Nature, rather than a member of society" (203) was to make an extreme claim for the primacy of nature and to imply that one could make a choice of allegiances. Thoreau recognizes a more complex reality near the end of the essay: "I live a sort of border life, on the confines of a world into which I make occasional and transient forays only" (243). The force of much of the essay, however, depends upon the polarities that he develops through his central metaphor of walking (including wilderness vs. civilization and Oregon and the West vs. Europe) and upon his provocative assertions of his own preferences: for the "odor of musquash," "tanned skin," "the impervious and quaking swamps." Thoreau's praise of swamps offers a good illustration of his rhetorical strategy and of the character of his defense of wild nature in "Walking." There is a calculated outrageousness in his challenge to substitute the swamp for the front yard ("Bring your sills up to the very edge of the swamp") and in the playful blasphemy of his declaration that "I enter the swamp as a sacred place, a *sanctum sanctorum*" (228). By presenting the swamp as the fertile place from which he derives his "subsistence" and where he refreshes and recreates himself, he identifies with the fundamental vitality of a nature seen as nourishing the spirit.

The preoccupation of critics with the tension between nature and society in "Walking," whether this tension is seen as productive or as limiting Thoreau's conception of wildness, responds to the importance of that tension in the essay.[67] Yet one can recognize that Thoreau habitually deploys wildness against civilization without concluding that this opposition is a sufficient explanation of his understanding of and attraction to wild nature. Thoreau's focus in "Walking" is on the symbolic implications of nature, as a stimulus to free and original and moral thought, rather than on the particularities that he renders in sometimes overwhelming detail in the journals. Yet the swamp could signify "the strength, the marrow of Nature" for him because he knew its biological richness from repeated observation. He gives the reader of "Walking" a glimpse of this richness by referring to "the dense beds of dwarf andromeda" and the "dull red bushes," sorted by name, that stand "in the quaking spagnum" (227). Thoreau uses wilderness symbolically in "Walking" to suggest the potential for self-discovery in "[pressing] forward incessantly," but even here he relies upon a visual shorthand to invoke the physical reality: "He would be climbing over the prostrate stems of primitive forest-trees" (226). It matters that Thoreau's experience of the Maine woods underlies such an image, whatever symbolic significance one might find in it. By appealing to "primitive" nature, Thoreau inevitably evokes its opposite, because this opposition is embedded in the language. He keeps returning to the particularity of this nature, however, as he continues to respond to new manifestations of wildness in his journals, because he sees it as representing a reality prior to and persisting despite human intervention.

Thoreau's other main theme in the essay, walking, provides a frame for his assertions about wildness and enables him to establish a convincing persona. He insists at the outset that such walking is demanding, taking one away from main roads on journeys of three to four hours, seeking not exercise but "the springs of life." It was spiritual exercise that Thoreau chiefly sought, and he regarded his walking as a form of meditation and a way of activating the senses. He makes the point that the environs of Concord are sufficient for such walking, insisting that he can find "an absolutely new prospect" any afternoon with a few hours walking. Exploring strange countries was not the object, rather opening the senses to the landscape at hand and finding in it the raw materials for reflection and self-examination.

Thoreau attempted to give walking a mythic dimension with his comparison of true walkers, or saunterers, with knights pursuing a crusade. Beneath the wittiness of his claims to embody the old "chivalric and heroic spirit" and to be engaged in reclaiming a holy land from the infidels, however, Thoreau makes an extreme claim for walking of the kind that he practiced. Although he does not mention Bunyan until later in the essay, he recalls Christian's pilgrimage in *The Pilgrim's Progress* indirectly by asserting, with mock fervor, that the walker needs to be "ready to leave father and mother and brother and sister, and wife and childe and friends, and never see them again" (206) and by invoking the Calvinist term "persevering."[68] Thoreau gives the Old Marlborough Road a symbolic character that suggests the New Testament way of faith that Christian follows ("It is a living way, / As the Christians say") but is more interested in showing how the imagination can travel anywhere on an abandoned road: "You may go round the world / by the Old Marlborough Road" (216). He invokes the Christian sense of life as a journey toward the New Jerusalem, as he does the language of Christian ritual elsewhere in the essay, to suggest the transformative character of walking and the kind of interaction with the natural world it involves. Thoreau redefines the Christian pilgrimage, whether to Canterbury or the Holy Land or heaven, as an inner journey whose objective was an invigorated understanding, a journey stimulated by the kind of sauntering he prescribes for his readers.

The fact that Thoreau habitually walked west matters, of course, since he associates the West with wildness and freedom. He could understand walking west as withdrawing from the city to the wilderness, seen as a virtually uninterrupted forest stretching to Oregon, and also as choosing the natural richness and promise of America over a depleted Europe. Thoreau instinctively identified with Columbus and with naturalists such as Michaux and Humboldt who reported the wonders of the primitive forests of the New World. In his admiring account of the panorama of the Mississippi with its steamboats and "rising cities" he may seem to embrace the popular understanding of westward expansion as progress, but Thoreau's West is also a mythic place, the "Great West" of the ancients and the inspiration for such fables as that of the gardens of the Hesperides. Comparing an American woodsman with Adam in paradise was a patriotic gesture but also an affirmation of the mythical understanding of America as a new Eden.

In the concluding pages of "Walking" Thoreau introduces a voice quite different from the challenging, prophetic one with which he begins the essay. This is a quieter, less assertive voice in which he recognizes the limits of knowledge and the transience of vision without abandoning his confidence in the quest that his walking embodies. Thoreau's observation that he lives "a sort of border life" is a way of suggesting the difficulty of following the kind of life that he calls "natural" and of truly knowing nature, "a personality so vast and universal that we have never seen one of her features" (242). It enables him to set up a distinction between the familiar, surveyed fields around Concord and "another land" in which the walker occasionally finds himself, thus preparing for the visionary mode of his accounts of the two sunset walks with which he concludes the essay.

In describing the first of these, on Spaulding's Farm, Thoreau establishes the physical scene of pinewoods with farmer's cart path leading through to a cranberry meadow, the whole illuminated by the rays of the setting sun. The vision that his imagination superimposes on this scene substitutes an idealized family in perfect harmony with the place for the "merely English nobility" that he rejected earlier. Aisles of pines form a "noble hall," lichen on the trunks of trees a coat of arms. Thoreau gives his natural nobility a leisure and serenity, and freedom from politics, that suggest the *otium* of pastoral poetry, but an *otium* that allows for vigorous intellectual activity. By rendering thinking as sound, like "the finest imaginable sweet musical hum" of distant bees, he preserves the mood of serenity. Thoreau shows the vision fading and recognizes the difficulty of recovering a sense of "cohabitancy" with such families, but the episode has the effect of establishing the importance of being able to summon such a vision of another world and of implying that it comes only to the walker, abroad and able to experience such moments as one in which a wood is transformed by the evening light.

Thoreau's story of climbing a white pine and bringing down the delicate red flowers he finds at the ends of the highest branches functions as another kind of exhortation to new ways of seeing and thinking. The failure of most to discover this unobserved beauty argues earthbound habits of mind. His insistence on living in the present, on hearing the cock crow in every barnyard, establishes a necessary precondition for breaking "rusty and antique" habits of thought. Thoreau's use of the image of the cock here recalls *Walden* (84), in which he announces his intention "to brag as lustily as chanticleer in the morning . . . if only to wake my neighbors up." In *Walden,* however, Thoreau identifies with the cock as a way of signaling his intent to challenge the habits and pieties of his neighbors. In "Walking" Thoreau includes himself among those called to attend to the present moment and forget their preoccupations. The cock's cry itself becomes a boisterous expression of the "health and soundness in Nature." The objective is still awakening, but Thoreau's emphasis at the end of "Walking" is on the importance of being ready to respond to the revelations of nature and to continue to seek what he calls in his last sentence "a great awakening light."

The fact that Thoreau describes himself as walking eastward, returning home, when he observes the "remarkable sunset" with which "Walking" concludes can be read as an acknowledgment of human limitation, perhaps even of mortality.[69] Thoreau recognizes that he has not reached his Holy Land, and it is apparent that the intensely realized scene that he presents is a transient one. Yet the lyrical celebration of possibility in this final section of the essay outweighs any sense of incompleteness or loss. Thoreau twice describes the "golden flood" of sunset light as "warm and serene." It makes a seeming paradise of the meadow through which he is walking and reassures with the promise of recurrence: "it would happen forever, an infinite number of evenings" (247). Thoreau's image of the sun on the backs of the walkers as "like a gentle herdsman driving us home at evening" recalls the quiet close of a Virgilian eclogue, in which the image of a shepherd returning his flock to the fold gives not only a sense of closure but assurance of an ordered world in which the cycle of activity will be repeated the next day.

Thoreau endows the light in this scene with a magical quality that it sometimes assumes in his writing, as in the "Spring" chapter of *Walden,* where he describes an early spring morning of "jumping from hummock to hummock" in the meadows "when the wild river valley and the woods were bathed in so pure and bright a light as would have waked the dead. . . . There needs no stronger proof of immortality" (317). This is an energizing light, suggesting the vigor of nature, as well as its freedom from any kind of human corruption. Although Thoreau uses the same adjectives in his description of the light at the end of "Walking," the mood is much calmer, reflecting a different time of the day and of the year (November): "We walked in so pure and bright a light, gilding the withered grass and leaves, so softly and serenely bright, I thought I had never bathed in such a golden flood, without a ripple or murmur to it" (247). One scene seems to kindle Thoreau's energies, the other to quiet them, but both suggest an escape into a timeless state and freedom from the pressures of society. Thoreau's emphasis on the softness and serenity of the light in the later scene gives nature a remarkable feeling of benevolence that carries over to the promise of transcendent vision in the concluding paragraph.

The image of sauntering toward the Holy Land and a "great awakening light" with which Thoreau ends "Walking" offers a revealing contrast with the triumphant conclusion of the first part of *The Pilgrim's Progress,* in which Christian and Hopeful, having crossed over the Jordan, ascend toward the perpetual light of the New Jerusalem with the help of the "shining Ones" who come out to greet them. In revising the goal of pilgrimage Thoreau offers his most dramatic challenge to Christian orthodoxy in "Walking." He expresses his desired vision as a light that would transform minds and hearts but that clearly belongs to the natural world we know ("as warm and serene and golden as on a bankside in autumn"), on this side of the river of death.

In several essays he wrote or revised near the end of his life ("Wild Apples," "Autumnal Tints," "Huckleberries") Thoreau moved away from the preoccupation with the nature of wildness and its liberating effects that one finds in "Walking" toward more intensive observation of natural phenomena.[70] Writing in an assured and often witty style, he draws upon his reading and his intimate familiarity with the countryside around Concord to show how seeking out and enjoying wild fruits and learning to see the colors of the autumn landscape can connect one with the vitality of the natural world. Of these essays, "Wild Apples" and "Huckleberries" reveal the most obvious connections with his earlier writing on wildness.[71] In "Wild Apples" the domesticated apple gone wild serves Thoreau as an emblem of the vigor and intensity of wild nature and as a vehicle for offering a view of human history and for exploring his own situation as a writer, by analogy with the irregular and stubbornly persistent trees.[72] Thoreau uses his quest for the wild apple to exemplify a way of living and thinking, appropriate to the field rather than the house, that he would have his readers embrace. The "Saunterer's Apple" must be eaten in its natural setting to be appreciated, preferably in October or November, "when the frosty weather nips your fingers . . . and the jay is heard screaming around."[73] Thoreau claims a moral right to such fruit, wherever found, for walkers, children as wild as the apples, and "the wild-eyed woman of the fields," a new heroic type who appears in "Huckleberries" as well.[74] The fatalistic note on which he concludes the essay ("The era of the Wild Apple will soon be past") reflects a growing concern with the erosion of wild nature that characterizes much of the late writing.[75] In the place of old orchards gone wild and naturalized apple trees scattered through the landscape, he sees a future of fenced-in plots of grafted trees and a corresponding loss of the pleasure of enjoying the "spirited and racy" taste of wild apples eaten in the woods and fields. By abruptly invoking Joel's diatribe, with its powerful evocation of the devastation of Israel's vineyards and fruit trees, to end the essay, Thoreau may have been playfully indulging his own tendency to exaggeration. Yet the effect of giving this Old Testament prophetic voice the space and prominence he did is to shift the essay into another register, invoking a moral framework that his audience could not ignore and giving his lamentation a deeper resonance: "The apple tree, even all the trees of the field, are withered: because joy is withered away from the sons of men."[76]

In "Huckleberries," drawn from his larger manuscript on wild fruits, Thoreau presents gathering and eating wild berries as a way of sustaining a "simple and wholesome relation to nature" of the sort enjoyed by his Indian precursors, whose use of berries he elaborately documents.[77] Nature spreads a table for "huckleberry people" as for animals, offering "a land flowing with milk and huckleberries."[78] With an "extra-vagance" that recalls "Walking," Thoreau portrays eating huckleberries as a form of communion affirming a sacramental relationship to nature: "We pluck and eat in remembrance of her."[79] What he celebrates in this essay, more than the huckleberry itself, is the "sense of freedom and spirit of adventure" he associates with going to the fields in search of wild berries. For Thoreau the "spirit of the huckleberry" or of any wild fruit is inseparable from the process of gathering it. To buy huckleberries from a cart or to look for one's apples in a barrel removes one from the shaping influences of nature that he sees as the source of the most important lessons. Hence the appropriation of the commons by landowners who claim the huckleberry fields becomes another sign of the kind of loss he laments in "Wild Apples": "The wild fruits of the earth disappear before civilization."[80] Thoreau made his fullest argument for preservation in response to this threat. His proposal that each town should preserve a primitive forest at its center is part of a larger argument for preserving natural features in their wild form: riverbanks, mountaintops, significant groves of trees, and outstanding specimens. For justification he appeals to the "higher uses" of "instruction and recreation" that he had invoked in arguing for national preserves in "Chesuncook," broadening his appeal here by noting how many people are "refreshed" by the "wild and primitive beauty" of the White Mountains.

Thoreau's "Autumnal Tints" differs from the other late essays in its focus on beauty and how we perceive it by an "intention of the eye" that disposes us to see what we are looking for. The essay is remarkable for its calm acceptance of mortality, understood as a natural ripeness; the falling leaves "teach us how to die."[81] Thoreau can identify with the purple stalks of the pokeberry, in its "perfect maturity" an emblem of successful life. The seasonal change with its subtle array of colors becomes for him nature's "annual festival," exciting "joy and exhilaration."[82] Thoreau still responds to what he perceives as nature's wildness, but with greater emphasis on aesthetic values. He takes pleasure in the "rich and wild beauty" of the deeply scalloped leaves of the scarlet oak, imagining it as like "some fair wild island in the ocean."[83] He celebrates the purples of the unmowed wild grasses ("the walker's harvest"), valuing them because the farmer does not. The tall, vividly colored Indian grass suggests a company of "red men" or an Indian chief, recurrent symbols of primitive America in Thoreau's writing. He still emphasizes the vitality of nature, but here its energy expresses itself as color. Bushes and leaves and grasses are "burning," "all aglow"; the red maple swamp is "all ablaze." The forest of bright leaves becomes his "flower-garden," eclipsing conventional gardens with their "few little asters and withered leaves."[84] While Thoreau continues to oppose the wild and the domestic, he seems more concerned with rendering the natural scene as richly and exactly as possible. He can now discriminate among wild grasses and is concerned with dating the succession of fall colors ("By the twenty-sixth of October the large scarlet oaks are in their prime").[85] The series of injunctions that concludes "Huckleberries" can be taken as a summary of Thoreau's views on how to live with wild nature, seen as both measure and source of health. Although the passage is incorporated from an earlier journal entry (for 23 August 1853) with relatively little revision, it takes on new force and poignancy as part of a manuscript on which Thoreau was working as his own health was obviously failing:

> Live in each season as it passes; breathe the air, drink the drink, taste the fruit, and resign yourself to the influences of each. . . . In August live on berries. . . . Be blown on by all the winds. Open all your pores and breathe in all her streams and oceans, at all seasons. Miasma and infection are from within, not without. . . . Grow green with spring—yellow and ripe with autumn. Drink of each season as a vial, a true panacea of all remedies mixed for your especial use. . . . For all nature is doing her best each moment to make us well. . . . Why nature is but another name for health.[86]

In urging that we "Live in each season as it passes" Thoreau was advocating that we attune ourselves to the rhythms of the natural world and allow these to act upon us. This was a way of experiencing wildness in familiar landscapes, shaped by human use but not entirely domesticated, that retained a vitality that could invigorate those able to appreciate it. Thoreau could find wildness in the pungency of domesticated apples gone wild and in huckleberries that flourished on lands burnt by fires set to clear brush, as well as in the many other kinds of wild fruit that he sought out.

The manuscript now available as *Wild Fruits* (2000), in an edition by Bradley Dean, provides some of the best evidence of Thoreau's belief that one could recover a primitive connection to wild nature without journeying to Audubon's Labrador or "Mackenzie's River," the kinds of remote places that kindled an appetite for wildness in his early reading of natural history.[87] Picking the berries of the shadbush near the Assabet River, Thoreau feels "as if I were in some remote, wild northern region where they are said to abound, maybe on the banks of the Saskatchewan."[88] Wild fruits, discovered as he ventures out from Concord, "feed the imagination" as "table fruits" such as exotic oranges or pineapples cannot.[89] Swamps in which he harvests cranberries become "little oases of wildness in the desert of our civilization" that lift us "out of the slime and film of our habitual life."[90] This opposition of the wild and the domestic runs through Thoreau's celebrations in *Wild Fruits* of the flavors and fragrances in which he delights, even the "strong invigorating aroma of green walnuts, astringent and *bracing* to the spirits, the fancy and the imagination, suggesting a tree that has its roots well in amid the bowels of Nature."[91] The cranberries that he finds in spring meadows possess "a refreshing, cheering, encouraging acid that literally put[s] the heart in you and set[s] you on edge for this world's experience."[92] In the last years of his life, however preoccupied he may have been with recording the habits of the plants he observed, Thoreau kept insisting upon the invigorating, transformative effects of recovering a relationship with what he saw as a primal nature persisting in the civilized landscape in which he chose to live.

3 Wilderness as Energy: John Muir's Sierra

*Nature is ever at work building &
pulling down, creating & destroying.
Keeping everything whirling &
flowing, allowing no rest but in rhyth-
mical motion, chasing everything in
endless song out of one beautiful form
into another.*

—JOHN MUIR PAPERS, 1912

John Muir found in Thoreau and Emerson mentors who pointed the
way to the discovery of spiritual truths in the natural world and
stimulated his appetite for exploring the wild nature he found in the
mountains of the Sierra Nevada when he arrived in California at the
age of thirty.[1] Thoreau became increasingly important to Muir as he
developed his own writing, yet one is immediately struck by the
differences in temperament and style that separate the two. Muir's
solitary, frequently dangerous expeditions in the mountains bear little resemblance to Thoreau's walks in the settled countryside
around Concord, and his rapturous celebrations of the divinity and
the vital force he found in the natural world express a kind of enthusiasm and desire to participate in the life of nature that one does not
find in Thoreau. Whatever Muir learned from Thoreau, one would
never confuse the two.

Muir read Thoreau avidly, with a thoroughness revealed by the
extensive notes and markings in his copies of Thoreau's works.[2] His
markings of passages in Thoreau's journals, all of which he read,

show Muir responding to accounts of animal behavior, botanical details, descriptions he particularly liked, observations on writing, and aphoristic comments on such topics as friendship and poverty, among other things. One can even see Muir using Thoreau's phrasing as a model for his own writing. In his endnotes to his 1868 copy of *A Week,* for example, he adapts a description by Thoreau of the freshness of the night wind blowing across a meadow in sketching his own description of an Alaskan glacier: "At night such a devout chaste icy freshness came down the Gl[acier] on the Sweet N[orth] wind." Muir's revisions of Thoreau typically reveal a more exuberant sensibility, expressed in affective language and a tendency to superlatives, as in his rewriting of Thoreau's "The impartial and unbribable beneficence of nature" as "The Eternal—everlasting immeasurable beneficence of Nature." What Muir loses in precision and subtlety, he gains in rhetorical force, at least for readers susceptible to the appeal of his enthusiasm.

While Muir deeply admired and learned from Thoreau, he carried on a sometimes chiding dialogue with him in his marginal notes. He resisted qualifications that might diminish the appeal of natural scenes. When Thoreau comments in "A Walk to Wachusett" that "distance and indistinctness lent a grandeur not their own" to mountains on the horizon, Muir objects: "Why not their own, have they not a right to all the air there is?" A comment in *A Week* that "the most stupendous scenery ceases to be sublime when it becomes distinct" prompts the defensive response: "no scenery is limited." Muir habitually explains landscape features described by Thoreau, such as parallel lakes in Maine, by invoking his knowledge of glaciation ("Direction of flow of the ice sheets"). When Thoreau describes Walden (in "The Ponds") as "a perennial spring . . . without any visible inlet or outlet except by the clouds and evaporation," Muir offers an explanation that dispels the mystery: "Walden is a moraine pond, wh[ich] dates back to the close of the Glacial period when the general New England ice sheet was re[c]eding & is fed by currents wh[ich] ooze thro beds of drift." The lecturing tone suggests the confidence of someone so absorbed in his studies of Yosemite glaciation that the scientific explanation seemed obvious.

Whatever his disagreements with Thoreau, Muir found much that engaged him in his reading and rereading of the works, including many of Thoreau's efforts to characterize nature's wildness. The categories in his endnotes to his 1868 edition of *The Maine Woods* include the pine tree, wolves, the loon ("the voice of the wilderness *enhanced*"), "wildness Indians and *forests*." He marked passages that evoke the wildness of North Twin Lake and that register Thoreau's sense of the indifference and the unfinished character of the nature he found near the summit of Katahdin.[3] Muir heavily marked the central section of "Walking," in which Thoreau argues the case for wildness, including paragraphs beginning "The West of which I speak is but another name for the Wild" and "In short, all good things are wild and free." His endnotes include "Wildness," "Bogs unfathomable," and "Wild in literature." Thoreau fed Muir's appetite for wildness and helped him to discover ways of thinking about it.

Muir would have found lessons in outrageousness in "Walking"—in the calculated provocation of Thoreau's tone and of his many claims for wildness, not to mention his parodies of religious language—but one can see him discovering his own ways of making "extreme" statements in his early prose. In an 1870 letter to his favorite correspondent, Mrs. Jeanne Carr, he proclaims the glory of King Sequoia as a way of justifying his unwillingness to come down from the Sierra range to San Francisco as she had urged. Unconstrained by the need to address a public audience, Muir poses as the mad prophet of a wilder religion than Thoreau could have imagined. The letter, written with Sequoia sap, reveals Muir at his most manic and playful:

> I'm in the woods, woods, woods, and they are in *me-ee-ee*. The King tree and I have sworn eternal love . . . and I've taken the sacrament with Douglas squirrel, drunk Sequoia wine, Sequoia blood, and with its rosy purple drops I am writing this woodsy gospel letter. . . . I wish I were so drunk and Sequoiacal that I could preach the green brown woods to all the juiceless world, descending from this divine wilderness like a John the Baptist,

eating Douglas squirrels and wild honey or wild anything, crying, Repent, for the Kingdom of Sequoia is at
hand![4]

The persona is an exaggerated version of the one Muir assumes in
much of his early prose, that of an enthusiastic, sometimes ecstatic
witness to the divinity of wild nature. Although the letter is characterized by a kind of gleeful abandonment that Muir would not allow
himself in print, it anticipates a fundamental difference between his
rhetorical strategy and Thoreau's. Muir was concerned with con-
verting his audience to his vision of wilderness glory, both to share
his good news and to recruit supporters in the battle for wilderness
preservation, in which he had become a major actor. His skills as a
storyteller and his ability to present an engaging persona—candid,
enthusiastic, adventurous to the point of recklessness—served his
rhetorical purposes.

Muir came to think of himself as a John the Baptist baptizing
his followers "in the beauty of God's mountains" and assumed the
role of prophet of a religion of wilderness as a national spokesman
for preservation in his later years.[5] He was less concerned with subverting conventional pieties, in the manner of Thoreau, than with
spreading his evangelical message by communicating his passion for
"pure wildness" and the intensity of his experiences of wilderness.
Encouraged by his new friend Robert Underwood Johnson, Muir
made the case for preservation with considerable effectiveness in
two influential articles in the monthly *Century* in 1890, urging the
creation of Yosemite National Park, and in subsequent articles collected in *Our National Parks* (1901). These articles reveal a practical-
ity and political awareness largely absent from his earlier writings,
but the passion Muir brought to these causes and his celebratory
style go back to the early years spent living mainly in Yosemite Valley. In exploring Muir's sense of wild nature, I will concentrate on
writings that draw upon the experience of those years, especially
My First Summer in the Sierra, revised in Muir's early seventies but
based upon a reconstruction in 1887 of his experiences in the summer of 1869.[6]

In his *Walden* fantasy about eating a woodchuck raw, Thoreau makes the appetite for wild meat a metaphor for the "primitive rank and savage" life he sometimes craved. In "Walking," even wilder food ("the marrow of koodoos devoured raw") suggests Thoreau's desire for a wildness so powerful that civilization could not endure its "glance." In his King Sequoia letter Muir plays with a similar opposition, between the wild food he found in the mountains and the insipid offerings of the lowlands:

> Douglas squirrel is so pervaded with rosin and burr juice his flesh can scarce be eaten even by mountaineers. No wonder he is so charged with magnetism! . . . I would eat him no matter how rosiny for the lightning he holds. I wish I could eat wilder things. Think of the grouse with balsam-scented crop stored with spruce buds, the wild sheep full of glacier meadow grass and daisies azure, and the bear burly and brown as Sequoia, eating pine-burrs and wasps' stings and all; then think of the soft light-ningless poultice-like pap reeking upon town tables.

Thoreau drew his examples in "Walking" from his reading (of Hottentots and their eating habits), Muir his from the experience of observing, and actually eating, the Douglas squirrel. When he longs for still wilder food, he seems to be thinking less of the grouse itself than of the spruce buds it eats, less of the bear than of its enviable ability to eat "pine-burrs and wasps' stings and all." Muir shows more interest in the pungent, gritty materiality of the wild nature he is unwilling to leave than in the idea of the primitive and savage. To become wilder, as Muir declares he will, implies a desire to survive on the resources of nature and absorb its flavor, as the Douglas squirrel and the grouse do. Muir could be playful about the benefits of a wild diet, as when he offers Sequoia juice as a remedy for "sick grangers and politicians" ("come suck Sequoia and be saved"), but beneath his exaggerations one sees a commitment that both recalls Thoreau and transforms Thoreau's sense of the nature of wildness and how to experience this. Muir would become like the squirrel or the grouse in their freedom and self-sufficiency in the wild, liberated from his dependence upon bread and the commitment to societal norms that it signified. He was resisting expectations that he would settle into a conventional vocation and finding ways of prolonging a period of exploring the Sierra that would stretch to 1875.

In his early years in California Muir sought a more absolute break with society than Thoreau, with his multiple ties to Concord, would have found appealing. He needed to provide himself with food and shelter and did this by managing the sawmill of the first inn in Yosemite Valley, but his accounts of his rambles in the mountains, fortified only with tea and bread, suggest that he aspired to the condition of the creatures he admired. In "A Near View of the High Sierra" he describes a night in a pine thicket that he shares with nesting birds in terms that suggest he would like to think of himself as able to survive in the same fashion that wild creatures did: "These are the best bed chambers the high mountains afford—snug as squirrel-nests, full of spicy odors, and with plenty of wind-played needles to sing one to sleep."[7] The fact that the biting cold forces him to creep out to the fire repeatedly is a reminder of his human limits. He could not be burdened with carrying blankets, but he obviously needed them.

Muir associates the wildness of the Douglas squirrel with the "magnetism" and "lightning" he finds in it; when one runs over his feet, he experiences "a thrill like a battery shock." In a draft of the chapter on the Douglas squirrel that he published in *The Mountains of California* (1894), Muir connects this electricity with the pine cones the squirrel eats: "his movements are perfect jets and flashes of energy as if surcharged with the refined fire and spice of the trees on which he feeds."[8] One of the fundamental aspects of wild nature for Muir, perhaps the most fundamental, was its energy. He observes this energy wherever he looks and seems charged by it himself, as though absorbing the power and excitement he senses around him. Thoreau asserted in "Walking" that "The most alive is the wildest" and found this wildness invigorating: "Not yet subdued to man, its presence refreshes him."[9] Muir's preoccupation with the energies of wild nature can be seen as another way of insisting upon its vitality and our need to be nourished and stimulated by this. He diverged from Thoreau, however, in how he understood this vitality and in the degree of importance he gave it. Thoreau tended to associate vigor in nature with a primitive state best embodied for him by swamps, where he found nature in its most fertile and least altered form. For Muir, the vitality of nature had more to do with an underlying principle of continuous change, which he found everywhere in the natural world: in the restless movements of the Douglas squirrel and the ouzel; in the rushing of mountain streams and the wilder cascades of Yosemite waterfalls, and, on a different timescale, in the movement of glaciers and the continual shaping of the Sierra landscape by the action of ice, water, and wind. He understood movement and change in nature as part of an ongoing process of creation, revealing the progressive unfolding of divine design. This flow, as he often called it, could be quick ("swirling") or barely perceptible, but it represented for Muir a kinetic force by which nature was "ever at work," creating and destroying, translating matter "out of one beautiful form into another."[10]

Muir's description of the view from the summit of Mt. Ritter demonstrates the way he saw the larger movements of nature as composing a divine harmony and offers another revealing comparison with Thoreau.[11] Anyone reading Thoreau's account of climbing Katahdin (in *The Maine Woods*) alongside Muir's rendering of his experience on Mt. Ritter (in "A Near View of the High Sierra," published in *The Mountains of California*) will be struck by the differences in their responses to what they perceive as wilderness. As we have seen, Thoreau emphasizes his deep uneasiness in the face of a landscape that he finds chaotic and intimidating in its "unfinished" state. This "vast and drear and inhuman" nature resists his efforts to find pattern and meaning in it.

Muir describes reaching the summit of Ritter after a "nerve-shaken" moment of paralysis on a cliff face, scrambling over the "savagely hacked and torn" surface of the mountain with new strength and confidence after miraculously recovering his presence of mind. He exults at having escaped from the "terrible shadow" of likely death into the "blessed light" of the summit, a light that "seemed all in all," as if charged with divinity. Looking to the south, beyond the wild spires of "the Minarets," he perceives "a sublime wilderness of mountains." His earlier perception, as he began to climb the difficult north face of the mountain, had been of a "wilderness of crumbling spires and battlements, built together in bewildering combinations" (*Mountains of California,* 44), a disorderly scene that aroused an uncharacteristic sense of foreboding. This is wilderness in its disorienting aspect. The tremendous views that Muir enjoys from the summit, exhilarated by his sense of deliverance, reveal a radically different understanding of wilderness, as a sublime landscape unified by "far-reaching harmonies." Thoreau's circumstances on Katahdin were of course drastically different from Muir's—laboring through boulder-strewn terrain that he can scarcely see in the clouds, on a mountain that bears little resemblance to Ritter. Yet even if one could imagine Thoreau in Muir's circumstances, and with Muir's mountaineering skills, it would be difficult to think of him sharing Muir's enthusiasm or his perception of sublime harmonies in the landscape of peaks and glaciers he saw.

The key to Muir's sense of the landscape as ordered and harmonious was the knowledge of glaciation that enabled him to understand the processes by which it had been sculpted. He sees clusters of peaks "fashioned like works of art—eloquent monuments of the ancient ice-rivers" (*Mountains of California,* 48). The aesthetic sense that informs Muir's vision of Sierra landscapes is visible in his habitual use of metaphors drawn from art—canyons as "glacial compositions," Cathedral Peak as "a temple of marvelous architecture"—yet the most remarkable aspect of the scene Muir describes is the sense of animation he gives it. The flank of a range flows away from its summits in "smooth undulations"; streams appear "bursting forth" from glaciers. Although the scene may appear motionless in the silence, "as if the work of creation were done," Muir finds "incessant motion and change" in such phenomena as glaciers imperceptibly grinding rocks beneath them and in the much slower action of lakes wearing away granite shores.[12] The "eternal flux of nature" constitutes a fundamental principle for Muir, explaining the formation of the landscape over time ("Ice changing to water, lakes to meadows, and mountains to plains") and grounding his belief in a nature that is alive and continually changing.[13]

A passage from near the end of *My First Summer* reveals how comprehensive the metaphor of nature as eternal flux could become for Muir. He collapses geological and ordinary human timescales to find change in all natural phenomena, from the apparently static (glaciers, rocks) to the obviously kinetic (a windstorm, an animal in motion), seeing evidence of divine purpose and benevolence in everything he observes:

> Contemplating the lace-like fabric of streams outspread over the mountains, we are reminded that everything is flowing—going somewhere, animals and so-called lifeless rocks as well as water. Thus the snow flows fast or slow in grand beauty-making glaciers and avalanches; the air in majestic floods carrying minerals, plant leaves, seeds, spores, with streams of music and fragrance; water streams carrying rocks both in solution and in the form of mud particles, sand, pebbles, and boulders. Rocks flow from volcanoes like water from springs, and animals flock together and flow in currents modified by stepping, leaping, gliding, flying, swimming, etc. While the stars go streaming through space pulsed on and on forever like blood globules in Nature's warm heart. [14]

Muir's efforts to perceive and find ways of rendering this flow generate much of the force of his writing. One feels him straining to connect with the intense energies he finds in the natural world, from the violence of the storms he seeks out to the "lightning" he finds in the Douglas squirrel. If Thoreau's ideal was a literature that could express nature by transplanting words to the page "with earth adhering to their roots," Muir's was a prose that could capture what he perceived as nature's joyous energies, typified by the mountain streams he delighted in:

> The snow on the high mountains is melting fast, and the streams are singing bank-full, swaying softly through the level meadows and bogs, quivering with spangles, swirling in pot-holes, resting in deep pools, leaping, shouting in wild, exulting energy over rough boulder dams, joyful, beautiful in all their forms. (*MFS*, 157)[15]

Muir learned to express his enthusiastic response to the dynamism he perceived through the repeated use of present participles in his descriptive prose ("singing . . . swaying . . . quivering").[16] Their effect is to recreate for the reader his sense of living intensely in the moment.

Muir's *My First Summer* offers a convenient framework for exploring the vision of wild nature that he developed in the years he spent based primarily in Yosemite, 1869 to 1875. In this day-by-day account of a summer one can see Muir developing characteristic ways of responding to the natural world that he would elaborate in subsequent essays, particularly those published in *The Mountains of California* and *The Yosemite* (1912). While the text purports to represent the experience of one summer, 1869, it draws upon the experiences of several summers and reflects the attitudes and editorial judgments of Muir in his seventies, as well as the responses of the younger Muir. Michael Cohen has argued that the old Muir revised the image of the young Muir presented by the text by giving more attention to the social forces threatening the Sierra wilderness and complicating the position of the young Muir, in the process sharpening the contrast between nature and civilization.[17] I am most struck by the writerly instincts of the older Muir apparent in his revisions of the three-volume "Sierra Journal" of 1887, itself a much-expanded version of the original 1869 journal that reflects Muir's efforts to find language that would represent the excitement and the transformative character of his early experience of the Sierra. In his 1910 revision of the 1887 "Sierra Journal" Muir eliminated redundancies, rewrote flat or diffuse passages, and improved the flow and the dramatic character of the narrative by shifting some sections and dropping passages that might seem digressive (e.g., an account of a shelter that he built for his employer Don Pedro and himself collapsing on them in the night). Muir generally revised to achieve greater economy, concreteness, and emotional intensity, sometimes rewriting substantial passages. The more lyrical passages tend to be the most heavily revised, with Muir dropping words such as "spiritual" and rewriting some of his earlier efforts to capture his sense of union with the natural world ("we flow and diffuse into the air and trees and streams and rocks"; "Nature like a fluid seems to drench and steep us throughout with newness of life").[18]

Although it registers only parts of Muir's experience, *My First Summer* offers a distilled version of characteristic responses from his Sierra years (1869–75). I will use examples from *My First Summer* and other writing on the Sierra to argue that Muir's sense of the energy and the constant flux of nature is the most distinctive and important element of his vision. This should be seen, however, as part of a complex of attitudes that contribute to his understanding of what constitutes wildness. Muir could celebrate the natural world, rapturously, because he understood it as an expression of divinity. The austere Protestant fundamentalism in which he had been raised by an increasingly evangelic and rigid father metamorphosed into a religion of nature influenced by Muir's readings in Emerson and Thoreau but given distinctive form by his own adventurous embrace of the Sierra. Like these mentors, whose pictures stood on the mantel in the family home in Martinez, Muir understood the natural world as the source of spiritual meaning. If nature "made sense as sacramental sign of spirit," as Catherine L. Albanese argues, Muir could feel free to praise the many wonders he found in it.[19] Mountains become temples and all natural phenomena "glorious," a word Muir became conscious of using so often in his prose that he resolved to edit himself severely. He viewed natural objects in the "pure" wilderness of the Sierra as "conductor[s] of divinity" by which he saw himself "filled with the Holy Ghost."[20] Muir's appropriation of religious language was easy and instinctive rather than ironic, as it often was for Thoreau.

Muir habitually found divine love as well as divine power and the ongoing work of creation in the natural world. While he was capable of discovering this love in virtually all natural phenomena, he frequently associated it with flowers, which he saw as softening the harshest mountain landscapes. Reflecting on the flowers he encounters high on Mt. Hoffman, Muir observes: "Strange and admirable it is that the more savage and chilly and storm-chafed the mountains, the finer the glow on their faces and the finer the plants they bear" (*MFS,* 152). Flowers, like the song of the ouzel in an otherwise forbidding canyon, become a sign of nature's benevo-lence.[21] Muir delighted in the paradox that nature could be both violent and tender, finding evidence of this in the flourishing of "storm-beaten sky gardens."[22] Coming upon one of these in which hail has broken off many spring flowers, he at first recoils at the "destructive harsh look" but then moralizes the scene: "Yet Nature loves her gardens, and all that we call destruction is creation when viewed in its true relations."[23]

Muir had to reconcile his belief in an ordered nature continu-ously shaped by a loving God with the abundant evidence he found of the destructive power of natural forces, including floods, earth-quakes, and windstorms. He did this by portraying the transforma-tion of the landscape as invariably beneficial. The apparent "confu-sion and ruin" produced by an earthquake yield eventually to new gardens and groves.[24] Broken limbs and blowdowns left by a wind-storm are the work of "Nature's forestry," regenerating and perfect-ing the forests. Discovering the path of a hurricane on Mt. Shasta, with thousands of uprooted pines, seems to strain Muir's faith in the beneficial effects of natural violence, but even in this instance he can appeal to the beauty of the Sierra forests as a whole.[25] Where oth-ers might see only violence, Muir found gentleness as well, and he relished the combination:

> How fiercely, devoutly wild is Nature in the midst of her beauty-loving tenderness!—painting lilies, watering them, caressing them with gentle hand, going from flower to flower like a gardener while building rock mountains and cloud mountains full of lightning and rain. (*MFS,* 133)

Muir's habit of personification, extended to the elements as well as to creatures and plants, helps to explain how he could believe in a sentient, loving nature and see this as compatible with a nature capable of appearing "savage" or violent. One can criticize Muir for sentimentalizing the natural world by his insistence upon humaniz-ing it, yet personification was a crucial way of expressing his extra-ordinary empathy with this world. It enabled him to assimilate all

natural phenomena to his optimistic view of a healthy nature undisturbed by human intrusions. Nothing seemed to shake his assurance that "in God's wildness lies the hope of the world—the great
fresh unblighted, unredeemed wilderness."[26] Unlike Thoreau,
whose formulation he revises here, Muir came to see wilderness as
wholly incompatible with civilization. The dichotomy was absolute
for him.

Muir's vision of a benevolent, joyous nature necessitated
significant omissions. We read little about predation or death,
although Muir found ways of explaining both.[27] His responsiveness
to the vitality of nature in *My First Summer* and elsewhere depends
upon excluding anything that would undermine this view. In his
entry for 2 July, after describing the sunny day as "making every particle of the crystal mountains throb and swirl and dance," he asserts
that there is "no stagnation, no death. Everything kept in joyful
rhythmic motion in the pulses of Nature's big heart" (*MFS,* 73).[28]
The Douglas squirrel seems so charged with energy, "a hot spark of
life," that Muir cannot imagine it "ever being weary or sick" (*MFS,*
68). Acknowledging sick or dying animals, even weariness, would
contradict Muir's emphasis on the vibrant life of nature.[29] In a journal entry for 1872 he acknowledges the fact of predation but marvels at how "neatly, secretly, decently" the killing was done: "I never
saw one drop of blood, one red stain on all this wilderness. Even
death is in harmony here."[30] Muir assumes a wilderness decorum by
which bones and torn fur are kept out of sight.

Muir's exclusion of sickness and death from his wilderness
descriptions is related to what may seem an obsessive concern with
the cleanness of "pure" wilderness and with its obverse, the
uncleanness of humans ("mankind alone is dirty" [*MFS,* 58]). In *My
First Summer* this human tendency is exemplified by his shepherd
companion Billy, with his pants comically encrusted with layers of
grease and debris, a "queer character" who seems out of place in the
wilderness to Muir (*MFS,* 129). He reacts with disgust rather than
amusement to the "degraded" Indians he encounters on the trail to
Mono Lake, "queer, hairy, muffled creatures" with dirty faces and
rabbit-skin robes. Muir's assertion that "nothing truly wild is
unclean" (*MFS,* 226), prompted by this encounter, suggests that

freedom from pollution by human habits as well as by human actions (the interference with natural processes represented by sheepherding, logging, and mining) was critical to his sense of the purity of wilderness.[31] Dirtiness correlates with other failings in these cases; Billy is insensible to his natural surroundings, and the Indians beg for whiskey and tobacco. Yet Muir found humans invariably unclean ("Man seems to be the only animal whose food soils him" [*MFS*, 79]), unlike the squirrels who keep themselves clean "in some mysterious way."

Muir's admiration of the way the Douglas squirrel can eat pine cones without getting sticky, like his broad claim that in the Sierra landscapes he knows "everything is perfectly clean and pure and full of divine lessons" (*MFS,* 157), reflects his need to find an absolute standard of purity in nature, as well as the strength of his desire to transcend the "dusty" world from which he emerged following the flocks of his employer Delaney. He found this purity in the "living" air and "champagne" waters of the mountains and in landscapes scoured by glaciers and washed clean by summer storms. Muir's reaction against extreme examples of dirtiness may reflect personal fastidiousness, yet his insistence on the cleanness of "unblighted" wilderness is inseparable from his sense of its divinity. Maintaining the purity of his religion of wilderness depended upon escaping and excluding, insofar as possible, the polluting influences of civilization.[32]

Muir presents himself as a reverential student of nature in *My First Summer,* spending long days absorbed in "observing, sketching, taking notes" (*MFS,* 220) in an effort to discover "divine lessons" in the landscape. On Yosemite's North Dome he appears "humbly prostrate . . . eager to offer self-denial and renunciation with eternal toil to learn any lesson in the divine manuscript" (*MFS,* 132). Muir adapted the old figure of the Book of Nature, given new life by Emerson, in imagining the mountains as scripture that he set about to read. Yet, as John Tallmadge argues, Muir understood the Book of Nature as "a living document that is continually being rewritten."[33] While Muir liked to use the metaphor of nature as the scripture of divine love, substituting this for the Scripture he was forced to memorize by his father, he subverted the metaphor by his instinct for finding and celebrating dynamic change in the natural world. He appears as much worshiper as student of nature in *My First Summer,* an ecstatic celebrant unable to contain the excitement that makes him want to cross the boundary between observer and observed and participate in the pulsing life around him.

Muir's reliance upon aesthetic terminology may appear to work against his sense of nature's dynamism at times, as when he sketches a "picturesque" scene, such as a mossy bridge over an Alaskan stream, or a "sublime" view of mountains or glaciers. His use of the term "picturesque," in particular, seems less a habit than a considered response to the tastes of readers of the essays he was sending to Eastern magazines.[34] In *My First Summer* Muir appeals to the aesthetic sense of his readers but typically energizes the scenes he describes by conveying a strong emotional response to them, frequently placing himself in the picture. He describes himself lying on an "altar boulder" in the middle of the North Fork of the Merced and recreates the scene with a discriminating eye and ear, evoking the mossy surface, the clear green water, a half-circle of lilies, "flowering dogwood and alder trees leaning over all in sun-sifted arches," and the mingled tones of the "water music" (*MFS,* 48–49). What animates the description, however, is the way Muir conveys his pleasure in the restful coolness and the interplay of light and sound. He moves, characteristically, from registering sensuously charged detail to spiritualizing the scene: "The place seemed holy, where one might hope to see God" (*MFS,* 49).

Muir evokes the beauty he finds in the Sierra landscape with a detailed descriptive language sensitive to color and texture and to nuances of sound, but he also registers its force on his senses. Beauty becomes experiential for him, something felt rather than merely appreciated by a detached observer. Describing the surrounding scene as he sketches on North Dome, Muir asserts: "The whole body seems to feel beauty when exposed to it as it feels the campfire or sunshine, entering not by the eyes alone, but equally through all one's flesh like radiant heat, making a passionate ecstatic pleasure-glow not explainable" (*MFS,* 131). He subsequently describes a rapturous experience in mountain meadows in similarly physical terms, attributing a kinetic force to the beauty he perceives: "In the midst of such beauty, pierced with its rays, one's body is all one tingling palate" (*MFS*, 153). By associating beauty with the influence of the sun's rays Muir makes it seem a power inherent in the natural world, independent of the sensibility of the viewer. It becomes another way in which the vitality of the Sierra landscape manifests itself.

In *The Mountains of California* Muir recalls sauntering through a glacier meadow feeling "dissolved" in a scene that gives him "inexpressible delight": "You are all eye, sifted through and through with light and beauty."[35] The image recalls Emerson's "transparent eyeball" (Muir carried Emerson's essays with him on some of his excursions in the Sierra) and suggests a self-effacing openness to the influence of the setting. Yet Muir more often represents himself as a "transparent body," in Sherman Paul's phrase, describing an "ecstatic pleasure-glow" or "tingling" that registers a transformation of his whole being that is more than a state of heightened perception, in which recognition of spiritual meanings becomes possible. In the entry for 6 June Muir describes his elation upon climbing out of the Sierra foothills:

> We are now in the mountains and they are in us, kindling enthusiasm, making every nerve quiver, filling every pore and cell of us. Our flesh-and-bone tabernacle seems transparent as glass to the beauty about us, as if truly an inseparable part of it, thrilling with the air and trees, streams and rocks, in the waves of the sun,—a part of all nature, neither old nor young, sick nor well, but immortal. (*MFS*, 16)

Muir can feel this beauty in "every nerve," as if registering its vibrations, because he is so ready to open himself to his surroundings and become "a part of all nature" in a more intensely physical way than were Emerson and Thoreau, from whom he borrowed the phrase. There is no sense of a forbidding or indifferent otherness here, as in Thoreau's account of his experience on Katahdin, only an eagerness to submit to the influence of the mountains.

Muir goes on in this entry to describe his experience as a "glorious conversion" releasing him from the "old bondage days" into "newness of life."³⁶ The "bondage" from which he feels released is most obviously that of a life spent pursuing a conventional vocation, but he also seems to feel released from ordinary human vulnerabilities such as sickness and age, even the passage of time, by identifying with the vitality he senses in the mountains. The "newness of life" he feels suggests the beginnings of a new sense of vocation, as "mountaineer," and also a sense of renewal that brings new energies. In a subsequent entry (13 June) Muir describes himself yielding to the rhythms of a Sierra day, "in which one seems to be dissolved and absorbed and sent pulsing onward we know not where" and describes this as "true freedom, a good practical sort of immortality" (*MFS*, 39). Immersing himself in the "pure wilderness" he has discovered, submitting to the flow of this landscape in motion, seems to promise freedom from social constraint and the need to make decisions about how to live. In a journal entry from 1875, Muir declares: "Every sense is satisfied. For me there is no past, no future—we live only in the present and are full."³⁷ This sense of living wholly in the present, and thus escaping the claims of a life bound by social and historical contexts, became a hallmark of Muir's renderings of his early experience in the Sierra.

The relationship to the Sierra landscape that Muir describes in *My First Summer* involves a reciprocity by which he draws energy and excitement, physical as well as spiritual, from a natural world that he perceives as pulsing with life and at the same time invests with his own sense of animation. Muir's nature "wears the color of [his] spirit," as Emerson would have put it, most obviously in his anthropomorphic treatment of animals, plants, streams, even rocks, which he frequently describes as sentient.³⁸ In an entry describing a July dawn, Muir asserts that "every pulse beats high, every living cell rejoices, the very rocks seem to thrill with life" (*MFS,* 124). The grand walls of Hetch Hetchy Valley, like those of Yosemite, seem to "glow with life."³⁹ Muir experienced an earthquake in which he saw rocks move, and he was well aware of the way boulders were transported by glacial action, but his treatment of rocks as living has more to do with his need to see all nature as animate and infused with spiritual life.⁴⁰ In an unpublished journal entry Muir notes:

"Compare walking on dead planks with walking on living rock where a distinct electric flash seems to attend each step."[41] His ability to feel such electricity speaks to his own sentience and to his conviction that he inhabited an unfinished, still-changing natural world.

The animals that Muir gave the greatest attention were those that displayed the most energy—in *My First Summer* the "small savage black ant"; the grasshopper; and his favorites, the ouzel and the Douglas squirrel, each given a chapter in *The Mountains of California*.[42] Muir romanticizes the ouzel, which he sees as the "poet" and "darling" of mountain streams, refining and spiritualizing their music in its sweet, invariably cheerful song. The ouzel becomes a kind of spirit of the waters for Muir, at home in the wildest streams and blending so perfectly with its environment that its mossy nest is virtually invisible. A part of the ouzel's appeal for him is its restless movement; he describes it "flitting about in the spray, diving in foaming eddies, whirling like a leaf among beaten foam-bells."[43] Muir sees the Douglas squirrel as even more energetic, "the wildest animal I ever saw,—a fiery, sputtering little bolt of life," stirring every grove with "wild life."[44] He romanticizes the Douglas squirrel as well, as the "wee hero" of the woods, but responds most strongly to its "lightning energy." The grasshopper becomes a more comic figure for Muir, providing a counterpoint to the "sublimity" of the mountains that he sketches from his perch on North Dome by its "glad, hilarious energy" (*MFS*, 139). His drawing of the grasshopper represents the path of its leaping, or what Muir calls its "singing" and "dancing," with its song rendered as "Hip, Hip, Hurrah." Muir's drawing of the Douglas squirrel in *My First Summer* shows it in several positions on a tree trunk, as if to suggest the way it jumps about observing him; in a drawing for *The Mountains of California* he represents the track of a squirrel running up and down a tree. If Muir's observations of these creatures, as of the Sierra landscape, clearly reflect his own moods, they also demonstrate the kind of kinetic energy that fed these moods. The leaping grasshopper invites description as a "crisp, electric spark of joy" (*MFS*, 141).

Muir's drawings of trees in *My First Summer* are more static, not surprisingly, and his botanical descriptions can be inert, yet his

imagination often endows the Sierra trees with a sense of animation. Pines in the wind become a "bright, waving, worshiping mul-
titude," and he imagines individual trees "throbbing, thrilling,
overflowing, full of life in every fiber and cell" (*MFS*, 52). The
juniper seems oddly resistant to his imagination ("I have ever found
it silent, cold, and rigid"), but Muir's attraction to other trees can
seem almost erotic.[45] On Mt. Hoffman he climbs a hemlock in
bloom, "with all its delicate feminine grace and beauty of form and
dress and behavior," to "revel in the midst of it" and enjoy the touch
of rich dark flowers that "make one's flesh tingle" (*MFS*, 151–52).
Muir's descriptions of storms acting upon forests, awakening them
to intense life, suggest the play of eros in nature. Clouds descend
into the trees "fondling their arrowy spires and soothing every
branch and leaf with gentleness in the midst of all the savage sound
and motion"; pines bend and toss their branches "as if interpreting
the very words of the storm while accepting its wildest onsets with
passionate exhilaration."[46]

   Muir saw the natural world as most vibrantly alive in storms
and other violent natural phenomena (floods, earthquakes,
avalanches) that we routinely label "disasters" because of their
destructive power but that he welcomed as part of a healthy process
of renewal. His habit of seeking out such events reflects an almost
compulsive desire to experience and understand the forces in play.
When Muir climbs a Douglas spruce for a view of the storm he
describes in "A Wind-Storm in the Forests," clinging to its swaying
top "like a bobolink on a reed," he participates in what he describes
as the "wild ecstasy" of the forest.[47] Muir's discriminating senses
register complex effects of light and sound and smell, all seen as
part of nature's "high festival." He perceives these effects as harmonious and as expressing a state of heightened excitement, a "wild
exuberance of light and motion," that mirrors his own. Such wildness—such a "rush and roar and ecstasy of motion," as he describes
it in a journal of an Alaskan trip—showed the natural world at its
most vigorous and irresistible.[48] Climbing the Douglas spruce was
Muir's way of immersing himself in this world and experiencing the
full force of the storm. He maintains his capacity to function as an
acutely perceptive observer but shows none of Thoreau's concern

with the "border" separating human and natural worlds, perferring to identify with the bobolink and with the trees themselves ("We all travel the milky way together, trees and men").[49]

Muir's description of an even more violent storm in "Flood Storm in the Sierra," published in *The Overland Monthly* in 1875 and revised for *The Mountains of California* (as "The River Floods"), suggests the relative value he placed on nature and human society ("True, some goods were destroyed, and a few rats and people were drowned, and some took cold on the house-tops and died, but the total loss was less than the gain").[50] The "gain" for Muir, apart from the excitement of experiencing the forces of nature at such a pitch of intensity, can be seen in the way he describes the storm awakening and animating the landscape. Characteristically, he finds "tones and gestures inexpressibly gentle manifested in the midst of what is called violence and fury." He sees the extraordinary rains as entirely benevolent in their effect on the landscape, bringing out new colors and fragrances in the parched woods: "I sauntered down through the dripping bushes, reveling in the universal vigor and freshness with which all the life around me was inspired. The woods were born again. How clean and unworn and immortal the world seemed to be!"[51] Muir seems to marvel here at nature's powers of recovery and the way the "born again" woods renew the promise of immortality he finds in the natural world. The adjective "unworn," which he also uses of a band of mountain sheep he sees at close range "looking as unworn and perfect as if created on the spot," suggests a nostalgia for a pristine natural world with the freshness of Eden.[52]

Muir found high excitement in Sierra streams and waterfalls at more normal times. In their movement and sound they expressed the joy he typically found in the natural world and offered the most visible evidence of the energies that kindled his enthusiasm. Muir celebrated the vigor of mountain streams and attributed to them the joy he felt in their presence, as in his description of "the glad, young Tamarack Creek, rejoicing, exulting, chanting, dancing in white, glowing, irised falls and cascades on the way to the Merced Cañon" (*MFS*, 102). The music of such streams, "river songs," appealed to him as much as their motion. He hears the Tuolumne

River in its canyon singing "the endless song of Creation shaking the devout listener into newness of life."[53] The song is endless because it suggests nature perpetually renewing itself, the listener devout because Muir assumes that the sanctity of such a place demands reverence. He invests the river song with an energizing, transforming power, invoking New Testament language ("newness of life"), here as elsewhere, to suggest an experience analogous to conversion. In his description of another stream, Cañon Creek, "falling, swirling, flashing from side to side in weariless exuberance of energy," one can see Muir's wonder at this inexhaustible energy, as well as a note of envy in someone reluctant to admit the need for sleep ("It seems extravagant to spend hours so precious in sleep" [*MFS*, 89]).

Waterfalls, concentrating the energy of streams, fascinated Muir by their power, as well as by the continuously changing spectacle they offered. He describes Nevada Falls as like "a living creature, full of the strength of the mountains and their huge, wild joy" (*MFS*, 188). Muir's published accounts of his dangerous approaches to Yosemite Falls emphasize their "nerve-straining" aspects; in *My First Summer* he shows himself yielding to a compulsion to work his way out on to a narrow ledge at the top of the upper falls to a point where he can look "into the heart of the snowy, chanting throng of comet-like streamers" (*MFS*, 120).[54] The unpublished journals offer more interesting glimpses of a Muir struggling to comprehend and respond to the mesmerizing spectacle of Yosemite Falls. He describes them at one point as "an endless revelation—mysterious, unreadable, immeasurable," yet holding him spellbound.[55] Early in his experience of Yosemite, during his first winter in the valley, he records a day when the falls were "in a terrible power" and he responded by shouting until he was "exhausted and sore with excitement," hardly aware whether he was "in or out of the body." Trying to capture a sense of the power he was responding to, he offers a description of the water roaring like ten thousand furies, probably influenced by his reading of Milton's *Paradise Lost,* then stops himself short ("But I speak after the manner of men"), recognizing that it is his imagination that invests the falls with a sense of discord and confusion. Instead, he tells himself, this is "rock music," focusing on the process by which the waters are deflected, whirled in eddies, mixed with air, and then "moved over the brink with songs that go farther into the substance of our being than was ever touched with man-made harmonies." One can see Muir schooling his imagination as he replaces his intitial, melodramatic image with one that implies divine agency. His habitual recourse was to spiritualize natural phenomena, as he does here by insisting that the waters "bear pure heaven in every note."[56]

While his efforts to capture the intensity of his responses to storms and waterfalls are more memorable and more distinctive, Muir does record quieter moments in which he finds nature calming. Such moments may be associated with a campsite that Muir finds particularly restful, the "altar" boulder in the North Fork of the Merced on which he sleeps soothed by water music, or a forest he enters at sunset in which he experiences "the deep peace of the solemn old woods" (MFS, 103). Glacier meadows, however, seem to offer the deepest sense of calm he finds, promising a respite from the emotional strain of many of his other experiences in the mountains. When they are not devastated by grazing sheep, these meadows seem Edenic to Muir, gardens planted and implicitly nurtured by God: "One might reasonably look to a wall of fire to fence such gardens" (MFS, 95). The allusion to the flaming sword that blocks the entrance to paradise after the expulsion of Adam and Eve suggests both the vulnerability of these unprotected meadows and their power to recall a state of primal innocence and harmony with God. Muir tends to oppose the beauty of the meadows to the "savage" summits around them, finding in their flowers a humanizing gentleness and also a fragility that contrasts with the permanence of rock. He is careful to distinguish these meadows from the "snipped, repressed appearances" of "pleasure-ground lawns," imagining them rejoicing "in pure wildness, blooming and fruiting in the vital light."[57] Muir says of one such sunny meadow, into which he emerges from dense woods: "I should like to live here always. It is so calm and withdrawn while open to the universe in communion with everything good" (MFS, 204).

Steven J. Holmes sees in Muir's attraction to glacier meadows, as expressed in an 1871 letter to Clinton Merriam, an extension of the desire to find a home in the natural world that he traces in Muir's early journals and letters. The meadow described in the letter to Merriam becomes a "mountain mansion" that offers a sense of refuge and of companionship with grasses and flowers that recalls Muir's early botanizing expeditions in Wisconsin and Ontario. Holmes documents a strain of domestic imagery in the young Muir's writing about nature that he uses to challenge traditional images of Muir that he sees as contributing to our sense of a dichotomy between wilderness and civilization. What most interests him is the way the young Muir could feel both "human connection *and* solitude, familiarity *and* adventure, domesticity *and* wildness."[58] My own interest is in the way ideas of wilderness and wildness assumed increasing importance in Muir's published writing, as he became more concerned about human impact upon the Sierra and other natural areas. I would argue that Muir's ability to find a sense of domestic comfort in what he thought of as wilderness need not be seen as undermining the dichotomy between wilderness and civilization that runs through much of this writing. Rather, it enabled him to identify with wild nature and to champion it.

Muir's capacity for feeling at home in the mountains and especially in the more intimate and sheltered spaces that he found in high meadows suggests that he could embrace wild nature because he found it familiar and welcoming, at least in some of its aspects. In his chapter on glacier meadows in *The Mountains of California* he describes the feeling of being "contained in one of Nature's most sacred chambers, withdrawn from the sterner influences of the mountains, secure from all intrusion, secure from yourself, free in the universal beauty."[59] Muir finds reminders of "the old home garden" in daisies and larkspurs, but the freedom and exhilaration he experiences depend upon his sense of being exposed to the influences of "pure nature" unmarred by the intrusions of human activity. The "fertile wilderness" of mountain meadows offered a lush beauty and a sense of security that he could associate with paradise. Although he recognized the seasonal changes that transformed these meadows and the slower process by which they would be filled in and eventually disappear, he eagerly embraced the illusion of timelessness that they offered.

Although Muir's tone is most often excited and celebratory in *My First Summer,* it registers a range of moods, including profound calm, growing disgust at the stupidity and destructiveness of the sheep, and revulsion from human habits and activities that threaten his sense of the purity and cleanness of wilderness. Despite the shifts of mood and the episodic nature of the narrative, one can find some kinds of progression in *My First Summer.* The story that emerges from journal entries beginning in early June and ending in late September traces Muir's travels from the lowlands to the high mountains and back and describes a spiritual journey that takes him from an initial sense of being transformed by his experience of the mountains through states of increasing emotional excitement as he explores the high places and, toward the end, to a growing regret as signs of the coming fall multiply. This new sense of the pressure of time, combined with the experience of returning to the hot and dusty lowlands after departing from the Tuolumne camp, suggest that it is not so easy to transcend the time-bound world as Muir's declarations about freedom and immortality suggest. By his closing prayer to return to the "Range of Light" he reaffirms his "conversion" and points to the direction his immediate future was to take, one that defied convention and expressed his desire to experience as much of the sense of timelessness and joy he found in the mountains as he could.

In journal entries that trace the progress of the summer Muir portrays himself as deepening his understanding of the nature of the Sierra environment and his sense of its fundamental vitality while coming to a new awareness of the unity of natural phenomena. A series of significant episodes near the end of Muir's summer offers a way to gauge his development: his first visits to Mt. Hoffman and Lake Tenaya (chap. 6); a night he spends at the top of Bloody Cañon (chap. 9); and experiences associated with his last campsite in Tuolumne meadows, including his first ascent of Cathedral Peak (chap. 10).[60]

In his entry for 26 July Muir describes a "boundless" day spent around the summit of Mt. Hoffman, the highest point he has reached to that time. The flowers he finds in this "storm-chafed" place strike him as "a cloud of witnesses to Nature's love in what we in our timid ignorance call a howling desert" (*MFS*, 153). They combine with sparkling minerals in the rocks to create a radiance that Muir sees as "a mirror reflecting the Creator." He transforms the site that others might view as "savage" by the sympathetic power of his imagination. In a particularly revealing paragraph Muir describes himself as so rapt in his absorption in the "sky gardens" of the summit that he loses any sense of time or direction:

> From garden to garden, ridge to ridge, I drifted enchanted, now on my knees gazing into the face of a daisy, now climbing again and again among the purple and azure flowers of the hemlocks, now down into the treasuries of the snow, or gazing afar over domes and peaks, lakes and woods, and the billowy glaciated fields of the upper Tuolumne, and trying to sketch them. (*MFS*, 153)

Muir's description of himself as drifting "enchanted" implies a complete relaxation of the will as he gives himself up to a moment-by-moment savoring of the place. Such drifting has none of the deliberateness of Thoreau's walking, which at least reveals a sense of direction and an alertness to analogies that enable new kinds of self-awareness. Rather, Muir seems to have entered a timeless, unreflective state in which he simply follows his impulses and saturates himself with the rich sensations that the setting affords. He gives the impression of being perfectly at home in the natural world, content to be directed by the stimuli to which he finds himself responding.

In his account of an excursion to Lake Tenaya the next day Muir describes a similar experience: "On I sauntered in freedom complete; body without weight as far as I was aware; now wading through starry parnassus bogs, now through gardens shoulder deep in larkspur and lillies, grasses and rushes, shaking off showers of dew"(*MFS*, 156).[61] Muir echoes Thoreau ("sauntered") while creating a sense of freedom from human constraints that is more extreme than anything one can find in "Walking" and showing himself unencumbered by Thoreau's sense of the limits of our ability to know the natural world. His assertion at the beginning of the entry, "We go home into the mountain's heart," reflects his growing level of intimacy and comfort with the place and perhaps his desire to find there a substitute for the conventional life he was resisting. Eventually he would find a home of a more familiar kind in Martinez, with his wife and his two daughters, but at this point the world below seems increasingly alien to him. Muir presents his early days in the mountains as an exhilarating time of discovery in a world where "everything is clean and pure and full of divine lessons," one of the more important of which is that anything we try to single out is "hitched to everything else in the universe" (*MFS,* 157).[62]

Muir's sense of belonging in the mountains and his faith in his ability to read their divine lessons are tested by his subsequent experience in Bloody Cañon on his excursion to Mono Lake. While he never admits to doubts, his description of a windy night spent in the "savage" canyon, after being importuned on his way there by the ragged band of natives he finds so dirty and repellent, suggests a tension between the potentially alienating effects of the place and his will to believe in the benevolence of nature. Muir describes himself as finding reassurance in the huge full moon rising above the canyon rim, recalling Wisconsin harvest moons, although in the process he reveals an underlying restlessness: "The effect was marvelously impressive and made me forget the Indians, the great black rocks above me, and the wild uproar of the winds and waters making their way down the huge jagged gorge. Of course I slept but little and gladly welcomed the dawn over the Mono Desert" (*MFS,* 222). Muir had previously described the uproar of wind and water as "making a glorious psalm of savage wildness" (*MFS,* 221). His use of the word "savage," here and in descriptions of the beauty of alpine flowers in juxtaposition to harsh-seeming rock faces, serves as an indicator of places where the imagination has to work harder to make wildness appealing and may not wholly succeed ("I slept but little").

The spiritual journey that Muir records in *My First Summer* reaches a climax in several entries from early September, including his account of finally climbing Cathedral Peak and discovering there the cassiope that has eluded him before, "blessed cassiope, ringing her thousands of sweet-toned bells" (*MFS*, 250).[63] The name Cathedral Peak invites an outpouring of religious imagery ("all the world seems a church and the mountains altars"), but Muir concludes the entry by celebrating wildness itself: "How wild everything is—wild as the sky and as pure" (*MFS*, 252). Muir's description of a landscape in which "everything is flowing," quoted previously, comes from his days at the Tuolumne camp, as does a related passage in which he moves beyond the idea of nature as flowing to explain the significance of its transformations:

> One is constantly reminded of the infinite lavishness and fertility of Nature—inexhaustible abundance amid what seems enormous waste. And yet when we look into any of her operations that lie within reach of our minds, we learn that no particle of her material is wasted or worn out. It is eternally flowing from use to use, beauty to yet higher beauty; and we soon cease to lament waste and death, and rather rejoice and exult in the imperishable, unspendable wealth of the universe, and faithfully watch and wait the reappearance of everything that melts and fades and dies about us, feeling sure that its next appearance will be better and more beautiful than the last. (*MFS*, 242–43)

In this vision of the perpetual transformation of matter, Muir finds purpose ("use") and beauty in any imaginable kind of change. Apparent waste, such as the destruction of trees by a windstorm, and death become part of a process by which an "inexhaustible" nature continually reconstitutes itself. Muir arrives at this recognition when the signs of fall have become unmistakable, acknowledging that things fade and die as he has not previously. His mood has shifted from exultation at feeling a part of a landscape that seemed immortal, and from relishing freedom from the pressure of time and the obligation to find a livelihood, to satisfaction at recognizing the cycles that sustain the larger natural world. The energy that Muir perceives in the rush of streams and other signs of dynamic life becomes the generative force by which nature creates and destroys, "chasing every material particle from form to form, ever changing, ever beautiful" (*MFS*, 238).

Muir claims at one point that the best lessons of the summer were those of the "unity and interrelation of all the features of the landscape revealed in general views" (*MFS*, 240). In a subsequent entry he underlines the point, relating his perception of the unity of the landscape to his understanding of how it was sculpted and concluding that "the features of the wildest landscape seem to be as harmoniously related as the features of a human face" (*MFS*, 254). This perception of harmony, like his recognition of the transformations by which a dynamic nature renewed itself, deepened his attraction to the "pure wilderness" of the Sierra. He would return, in the following January, to spend another five years pursuing his studies in glaciation, a passion that would drive his later trips to Alaska, and rambling through the mountains to absorb as much as he could of the beauty and vitality he found there. What he managed to capture in *My First Summer,* rewriting the impressions he had recorded in his 1869 journal, was a progress from initial excitement to a deepened sense of intimacy with a place he regarded as embodying wildness: "More and more . . . we feel ourselves part of wild Nature, kin to everything" (*MFS*, 243).

In *My First Summer* and much of his early writing on the Sierra Muir offers a vision of wilderness that holds out the prospect of spiritual renewal and sensuous delight. While he recognizes past damage and future threats to this wilderness, his vision is largely ahistorical, ignoring, among other things, the fact that the Yosemite Valley he knew owed its parklike character to repeated burning by the native inhabitants of the valley, who had been evicted by troops in 1851.[64] Muir suggests that his readers can immerse themselves in the dynamic flow of the world he evokes and thereby achieve the "true freedom" and "good practical sort of immortality" that he found in his Sierra days. His celebratory and idealizing mode leaves little room for recognizing any emotion other than joy and, at its most rapturous, any time other than the present. The defense of "pure" wilderness in America has never been more confident or more absolute.

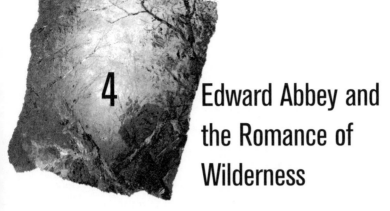

4 Edward Abbey and the Romance of Wilderness

Us nature mystics got to stick together.

——EDWARD ABBEY, *A Voice Crying in
the Wilderness*

At the end of his introduction to *The Journey Home* (1977) Edward
Abbey comments that if he sounds intransigent it is not just because
he loves an argument, and likes to provoke people, but because he
is an extremist, "one who lives and loves by choice far out on the
very verge of things, on the edge of the abyss, where this world falls
off into the depths of another" (xiv). The image, drawn from the
canyon country that Abbey claimed as his spiritual home, is a particularly revealing one. Anyone who knows Abbey's writing will
recognize one kind of extremism in his assaults on proprieties of
any kind he can imagine and his fantasies of subverting the ranchers,
developers, dam builders, and other enemies of wilderness whose
ways he never tired of ridiculing. Yet the image also suggests another
kind of extremism that Abbey recognized in himself—his attraction
to another, more elusive world that he found in the desert landscape; a "world beyond," as he often calls it. Abbey represents this
world as mysterious, boundless, ultimately unknowable, immensely
alluring and at the same time dangerous. I am interested primarily
in this kind of extremism and its implications for his writing.

   As others have observed, Abbey was a romantic despite himself.[1] His complicated efforts to deal with the romance of the desert wildness that he found so fascinating give his prose a tension and unpredictability that keep his readers off balance.[2] Abbey frequently suggests the allure and power of wilderness by appealing to the nonrational and the mysterious, yet he typically undercuts such appeals by offering an ironic gesture or returning abruptly to the concrete, physical world in which he locates himself. One can see the wary self-consciousness that generates this tension in Abbey's habit of calling his romantic impulses sick, even as he indulges them, as if he can authorize such indulgence only by showing that it is an aberration from the tough rationality with which he confronts the world. When Abbey allowed himself to become more lyrical and introspective in describing his experiences of wilderness, he found ways to signal his awareness of the limitations of his language.[3] The most remarkable thing about Abbey's portrayal of wilderness, however, is the way he continued to pursue its elusive and undefinable qualities while constructing defenses against the romantic tendencies he recognized in himself, relying upon irony, irreverence, and assertions of fact or rationality.

Abbey understood the impossibility of capturing the essence of the desert in language, as Claire Lawrence argues in a recent essay in which she reads *Desert Solitaire* (1968) as "the story of the failure of language to represent the real," finding a recognition of this failure in Abbey's statement near the end of the book that the desert "presents a riddle which has no answer, and that the riddle itself is an illusion created by some limitation or exaggeration of the displaced human consciousness."[4] This passage displays the self-awareness about the problems of representation that some recent critics have taken as a warrant for recasting Abbey as a postmodernist. It is important to note, however, that such a recognition is not sufficient for Abbey. He proceeds to question his conclusion about the desert as riddle: "This is at least what I tell myself when I fix my attention on what is rational, sensible, and realistic, believing that I have overcome at last the gallant infirmity of the soul called romance."[5] He can think himself cured, Abbey says, only as long as he stays away from the desert, recognizing his continuing susceptibility to its allure and mystery.

The most revealing of Abbey's many reasons for celebrating and defending wilderness, and the most relevant to a discussion of his romanticism, have to do in some way with transcending limitations. Not all of these reflect the kind of tension I have described. Abbey unproblematically praises wilderness as a place of liberty, as a source of both an exhilarating sense of personal freedom of the sort that he experiences as he and Newcomb begin the float through Glen Canyon described in *Desert Solitaire* and the political freedom his outlaw heroes (Jack Burns, Hayduke) seek and sometimes find there. He embraces the desert as a place that offers possibilities for physical trial and for vision, although he finds seeking visions more problematic than testing himself against the harshness of the country. Abbey can find Eden in this wilderness, in an unspoiled Glen Canyon, and he can imagine a posturban world in which sand dunes will have buried the cities of the desert Southwest. Both environments offer release from the constraints and oppressions of industrial society and the opportunity to live a more primal kind of existence consistent with what Abbey calls "our basic heritage of a million years of hunting, gathering, wandering."[6] Perhaps the deepest appeal of the desert for Abbey was as a place where boundaries are dissolved, those of time as well as those of space. The sense of mystery that he found there had to do with the spare and haunting beauty he perceived but even more with the way the desert seemed to open out in space and time, offering glimpses of a deeper and more elemental world.

The desert could become an image of desire for Abbey, it has been argued, because he shuts out "cultural and social complexities" such as those presented by nearby Navajo and Ute reservations, contemporary political conflicts, and the hazard of fallout from nuclear testing.[7] One could argue that other kinds of social realities impinge on Abbey's vision of the desert (Glen Canyon Dam, developers with their complaints of water shortages, the incursions of "industrial tourism") and that his references to "strontium" in the bones and radium damage to uranium miners and his inclusion of a story ("Rocks") illustrating the greed and violence unleashed by the scramble to find uranium register significant concern with the impact of the nuclear age on the Southwest. Yet Abbey seems less interested, ultimately, in cultural complexity and even in the pressure of culture on the desert than in the desert's capacity to assimilate human presence. Traces of this presence—in petroglyphs and Anasazi ruins, steps cut in the rock by Mormon pioneers, ghost camps of miners—engage him more than present social realities.[8] Such traces are absorbed by a landscape that Abbey sees as characterized by emptiness, silence, and a feeling of timelessness. He saved his account of the part of this landscape least marked by human presence for his penultimate chapter ("Terra Incognita: Into the Maze") and then emphasized its mysteriousness: "we have seen only a tiny corner of the Maze . . . the heart of it remains unknown" (DS, 293). It is worth asking what is lost by opposing wilderness and culture, but it is also important to ask what is gained by Abbey's effort to render the otherness he found in the desert landscape.[9]

Abbey's preoccupation with the mysterious and unknown is most apparent in *Desert Solitaire*. In this early work, with its focus on solitary encounters with wilderness, Abbey seems particularly concerned with exploring the possibilities and limitations of a romantic attraction to the desert. One finds him taking a more detached and ironic view of mystical impulses, his own and those of others, in some of his subsequent essays, particularly those from the decade after the publication of *Desert Solitaire* that he collected in *Abbey's Road* (1979). Abbey became increasingly preoccupied with the need to defend wilderness, offering his most radical pronouncements in the essays from the last decade of his life. He argues in his introduction ("Forward") to Dave Foreman's field guide to monkeywrenching that if wilderness is our true home and is threatened by destruction, both of which propositions he believed to be true, then "we have the right to defend that home . . . by whatever means are necessary."[10] He ends the essay—titled "Eco-Defense" when it was collected in *One Life at a Time* (1987)—by encouraging his readers to resist the "chainsaw-massacre" of public forests by spiking trees. Yet if Abbey's radicalism became more overt in the course of his career, his fascination with the otherness of the desert and the difficulties of trying to know this persists throughout his published work. Many of the motifs of *Desert Solitaire* recur in the later writing, and the idiosyncratic blend of toughness and romanticism that he achieved in that work would become his signature.

When an interviewer pressed him about whether he was a romantic, Abbey grumbled about the question but allowed that he considered himself one, because he was an idealist and because he succumbed to the "romantic ills" of thinking things must be more beautiful beyond the next hill and of being fascinated with "the mysterious and unknown."[11] In his writing, however, Abbey tends to undercut his more romantic imaginings, letting them play out and then checking himself as if unwilling to lose his footing in the tangible world. In an essay on Death Valley he describes himself looking into shadowy, unexplored canyons and feeling again "the old sick romantic urge to fade away into these mountains, to disappear, to merge and meld with the ultimate, the unnameable," then ends the paragraph with, "Easy now. What's the hurry? I light a cigar instead."[12] In another essay he recalls exploring out-of-the-way desert places, looking for the "back of beyond," in his student days at the University of New Mexico: "We were desert mystics, my few friends and I, the kind who read maps as others read their holy books. I once sat on the rim of a mesa above the Rio Grande for three days and nights, trying to have a vision. I got hungry and saw God in the form of a beef pie."[13] Abbey's joking about himself as mystic—"Us nature mystics got to stick together"—is a way of legitimizing a fascination with the world beyond the rim.[14] He could recognize in himself the mystic's attraction to the unknown and the infinite. Abbey did not confuse this unknown with God, although he found it hard to avoid talking about God in trying to describe the kind of mystery that attracted him. Thus on the river trip through Glen Canyon he describes in *Desert Solitaire* he asks, after hiking by himself into the canyon of the Escalante, "Is this at last the *locus Dei?*" He sees temples and cathedrals in the rock, then faults the weakness of imagination that leads us into "fantasies of the supernal." We should learn instead "to perceive in water, leaves and silence more than sufficient of the absolute and marvelous" (*DS,* 200). What is remarkable here is not so much Abbey's disavowal of conventional religious imagery and his affirmation of the visible reality around him—what he calls elsewhere in *Desert Solitaire* "the tangible, dogmatically real earth on which we stand" (190)—as his need to find the "absolute" and the "marvellous" in this place.

The canyon of the Escalante invites such speculations because Abbey experiences it as haunting, vibrantly alive, and mysteriously beautiful. It is transformed by a "radiant golden light"; the "crystal water" shimmers as it flows toward him; "mysterious and inviting" side canyons offer an occasional glimpse of the "delicious magical green" of a young cottonwood with leaves that vibrate "like spangles in the vivid air." Abbey's diction ("golden," "crystal," "spangles") elevates the scene, making it seem dreamlike and paradisal. Interestingly, the sonnet Abbey contributed to a collection of the letters of Everett Ruess conjures up a similar sense of the marvelous and alluring. It begins, "You walked into the radiance of death / through passageways of stillness, stone, and light, / gold coin of cottonwoods, the spangled shade." Like many others, Abbey was taken by the legend of Ruess, who disappeared in 1934 from the Escalante area in the course of his private pursuit of the mysteries of the desert—or, as Abbey puts it in his sonnet, of "the source, deep in the core, the maze, / the secret center where there are no bounds."[15] Abbey could offer this romantic vision of Ruess's disappearance because he knew the appeal of the mysterious and the boundless and could identify with someone who wandered the desert Southwest in search of a truth he could not find in cities. He also knew how to bring himself and his readers back to the mundane. Abbey ends the account of his experience of the Escalante by describing the long trudge back to his campsite in the dark and his late catfish dinner with Newcomb. Still, he leaves us with a sense of the holiness of the place by describing the whispering of leaves in the cottonwoods as "like the whispering of ghosts in an ancient, sacrosanct, condemned cathedral," falling back on religious language despite his disavowals of the supernatural. Earlier we see him playing hymns from his childhood on the harmonica ("We shall gather by the river") as he and Newcomb drift deeper into Glen Canyon.

Abbey seems close to disappearing into the landscape himself in the Havasu chapter of *Desert Solitaire,* first when he describes himself slipping into a state in which he dreams away days in his private "Eden" on Havasu Creek and then in his brush with death when he finds himself trapped in a side canyon that initially defeats all his efforts to escape. The chapter can be read as a commentary on the hazards of prolonged solitude and the absorption in the sensuous appeal of the natural world that it invites. With mild irony, Abbey describes himself worrying about butterflies and "who was dreaming what" and slipping into "lunacy" as he looks at his hand and sees "a leaf trembling on a branch" (*DS*, 225). He conveys the danger as well as the seductiveness of these "wild, strange, ambiguous" days, as he shows himself losing his sense of the boundary that separates him from the natural world. Abbey presents his adventure in the canyon as a corrective to this romantic reverie, an experience by which he "regained everything that seemed to be ebbing away" (*DS*, 226)—presumably consciousness of his own individuality and of the claims of a world outside that of Havasu Canyon. What he learns from his near-fatal hike is that the natural world can be sinister and "treacherous," as well as alluring—that it can easily kill him, in fact—and that he cares intensely about living. The images with which Abbey ends the chapter—bellowing the "Ode to Joy" as he hikes back, happily sheltering for the night in a coyote den—displace those of languorous days of dreaming by the creek and express a vigorous affirmation of life. Abbey's Havasu experience, from a trip taken when he was twenty-three and still a student, offered a good story and also a vehicle for revealing a particular kind of romantic excess, in a book preoccupied with testing how far one can go out on the edge of awareness of the natural world without losing a sense of one's individual, human identity.

Abbey tells his readers at the beginning of *Desert Solitaire* that he wants to "look at and into" the juniper and the quartz, the vulture and the spider, to confront "the bare bones of existence, the elemental and fundamental, the bedrock which sustains us"(6), without relying on personification or other learned ways of perceiving the natural world. He wants to recover a capacity for direct, unmediated vision and to bridge the gap between the human and the nonhuman: "I dream of a hard and brutal mysticism in which the naked self merges with a nonhuman world and yet somehow survives still intact, individual, separate. Paradox and bedrock" (*DS*, 6).

Abbey uses the term "paradox" here to refer to the struggle to hold onto a sense of separateness while seeking some kind of union with the desert world.[16] Christian mysticism would not allow this kind of paradox, by which the self retains its individuality, however naked, rather than dissolving into the divine other. Nor could it be called "hard and brutal." The mysticism Abbey seems to be reaching for here has more in common with the "harder mysticism" that Robinson Jeffers opposes to that of the Asian friend he addresses in "Credo"—what he describes as the "massive / Mysticism of stone" in "Rock and Hawk"—than with anything in Christian or eastern mystical traditions.[17] By "hard and brutal" Abbey implies a commitment to the "dogmatically real" world of the desert comparable to Jeffers's commitment to the world of Big Sur, with its granite headlands marked by "ages of storms." Both writers declare their allegiance to the tangible, enduring reality of their chosen places, rejecting the possibility of better worlds elsewhere.

At one point in *Desert Solitaire* Abbey speculates that perhaps "thought is an illusion and only rock is real. Rock and sun" (219). Speculation of this sort, typical of the contemplative mood that Abbey often slips into in *Desert Solitaire,* is part of an effort to disengage from habits of "human qualification." The "dogmatic clarity" of the desert sun dissolves the explanations of religion and philosophy for Abbey. He insists that the desert says nothing and means nothing: "the desert lies there like the bare skeleton of Being, spare, austere, utterly worthless, inviting not love but contemplation" (*DS,* 270). An attraction to the austerity and harshness of the desert runs through Abbey's work, but he seems particularly taken by this in *Desert Solitaire.* He describes sitting in the heat of July, at the "crucial hour" of noon, when "the desert reveals itself nakedly and cruelly, with no meaning but its own existence" (*DS,* 155). In this aspect the desert seems particularly unaccommodating and resistant to interpretation, yet Abbey tells us that he contemplates the scene of juniper and bedrock and prays, "in [his] fashion, for a vision of truth" (*DS,* 155). One of his preoccupations in *Desert Solitaire* is with the paradox that for all its "clarity and simplicity," the desert wears a "veil of mystery," hinting at something "unknown, unknowable, about to be revealed" (*DS,* 271). Abbey will declare that the desert has no meaning but itself and then look for revelations. The combativeness with which he defends his commitment to "the surface of things" ("what else is there? what else do we need?") seems in part a compensation for his own need to look for something beyond the surface of the scenes he renders with such particularity.[18]

It is tempting to think of Abbey as a kind of secular desert father who withdraws into the wilderness to seek visions and find the prophetic voice with which to deliver his jeremiads against industrial society. Abbey fantasized about living by himself in "a stone hut deep in the desert . . . wandering about naked, reading, writing, thinking, playing flute, dreaming, doing nothing at all; simply being" (Conf., 258). Mortifying the flesh is not a part of this fantasy, of course, nor is finding God. The analogy with the monastic life of the early Christian desert fathers breaks down quickly. For one thing, Abbey liked the companionship of his friends, and his women, too much to disappear into the desert for very long. For another, he preferred a different kind of image of himself, as "desert rat." When he describes the harshness of desert landscapes in the essays that followed Desert Solitaire, part of the point is to show that he can survive and even enjoy them. He identifies with the "old time desert rats" who liked life in the desert because it was "gaunt and spare" (BW, 86). In his journal Abbey embraced the "indifference" and even "hostility" he found in the natural world: "I like it tough" (Conf., 116). In his essay "How it Was" he summarizes conventional answers to the question "Why wilderness?"—all of which he endorses—then supplies a more personal one: "because we like the taste of freedom; because we like the smell of danger" (BW, 59).[19] Wilderness appealed to Abbey as a place where he could escape from the constraints and preoccupations of the everyday world and test himself against whatever hazards nature might present, whether on wild rivers or in harsh and often unfamiliar desert landscapes.

Abbey ventured into unforgiving desert terrain over and over again, exploring blind canyons; undertaking a punishing solo trek in the Cabieza Prieta relatively late in his life (described in "A Walk in the Desert Hills"); seeking out the most desolate landscapes he could find in trips to Death Valley and the Pinacate region on the Gulf of California, which he regarded as the "ultimate wasteland"

and "the final test of desert rathood" (*BW*, 151). Abbey begins his essay on the Pinacate with an homage to the desert rat as capable of loving country most people find unlovable and "tolerably adapted to intense heat, constant glare, sand in his eggs, scorpions in his shoes, kissing bugs in his bedroll" (*BW*, 150). This is the vein Abbey works with a kind of mad glee in "The Great American Desert," where he assaults the reader with catalogues of hostile flora and fauna, along with tales of sunstroke and poisonous water. The point of all this seems to be to discourage the unworthy and to demonstrate his own toughness, with Abbeyish humor: "I approach nature with a certain surly ill-will, daring Her to make trouble" (*JH*, 18). Yet it also allows Abbey to get away with the romantic gesture with which he ends that essay, answering the question "Why go into the desert?" by showing himself looking into the distance, in the direction indicated by a mysterious stone arrow, and finding "nothing out there. Nothing at all. Nothing but the desert. Nothing but the silent world" (*JH*, 22).

Abbey's kind of desert rat, for all the demonstrations of toughness, is after more than freedom and adventure and the chance to prove his manhood. The ultimate appeal is that of the unknown and unknowable. Abbey says that "the desert rat loves the desert because there is something about it that he cannot explain or even name" (*BW*, 154) and finds the Pinacate inviting because it represents the desert in its purest form, without mountains or canyons, only a "vast desolate nothingness." He appropriates a line from Beckett, "Nothing is more real than nothing," to explain the "great brooding solemnity" in an ocean of sand dunes that he finds in one area of the Pinacate (*BW*, 152). Abbey keeps coming back to the idea of nothingness in his meditations on desert places. In *Desert Solitaire* he describes the silence he experiences when he climbs out of the canyon as taking on a "deeper dimension": "The sound of nothingness?" (208).

Abbey came as close as he ever did to explaining what he meant by nothingness in "A Walk in the Desert": "Land of *nada*, kingdom of *nihilo*. God knows there's plenty of both out here. But it's a positive nothingness, as an idealist would say, rich in time, space, silence, light, darkness, the fullness of pure being" (*BW*, 46).

This goes a step beyond the desire to confront the "bare skeleton of Being" that one finds in *Desert Solitaire,* to a richer sense of the vital presence of the desert. Where a Christian mystic might find in silence and vacancy a way to the apprehension of God, Abbey here mocks the notion of finding God in the desert. Looking out over the "vast, waiting, listening openness of the desert" from the campsite in the Australian outback that he describes in another essay, he exults, "*Gloria in excelsis nihilo!*" (*AR,* 66) But if not God, what? Something Abbey represents as stripped of human associations and projections, something elemental and inviting that he describes in his journal as "the bare incomprehensible *is-ness* of being" (*Conf.,* 185). He must fill the vacancy, the *nihilo* that denies the existence of God, with a sense of the "fullness of pure being."

Sometimes, particularly in his later work, Abbey appears to settle for simply participating in the "being" of the desert. In a journal entry he claims that the "present moment, fully lived, is . . . the only eternity we can know" (*Conf.,* 266). After speculating on the unfathomable secrets of petroglyphs in one essay, Abbey concludes that their message is "*We were here*" and accepts this as sufficient, recognizing the impossibility of saying what the desert means: "It means what it is. It is there, it will be there when we are gone. But for a while we living things—men, women, birds, that coyote howling far off on yonder stony ridge—we were a part of it all. That should be enough" (*BW,* 94). In much of his writing Abbey seems intent on seizing the present moment by recording his sense impressions in detail, in the process conveying his pleasure in the scene. He found various ways of questioning efforts to read further meaning into the visible world. His dialogue with Thoreau in "Down the River with Henry Thoreau," for example, prompts him to challenge the habit of reading nature as symbolic of a transcendent spiritual reality. The fading light has no meaning but its own intrinsic beauty, he asserts as if talking back to Thoreau, and "the planets signify nothing but themselves" (*Down the River,* 19—20). Camped out by himself on the rim of Cape Solitude, Abbey self-consciously performs his own kind of religious ritual, building a fire of juniper twigs ("no incense finer, no ceremony more fitting") and playing his flute, naked, in an act of homage to the place ("a song for the river and the great canyon, a song for the sky, a song for the set-ting sun"). Yet the desert world responds with "grand indifference." Abbey raises the possibility of dramatic revelations of divinity—God speaking from a cloud, an angel in "an aura of blue flame" float-ing along the rim toward him—only to scorn them as showing the kind of poverty of imagination he finds in the mysticism of Carlos Castaneda. The true magic, he insists, "inheres in the ordinary, the commonplace, the everyday, the mystery of the obvious" (*AR,* 195). Mystery is a better word than God, Abbey wrote in his journal, "because it suggests questions, not answers" (*Conf.,* 254).

A number of the essays collected in *Abbey's Road,* published a decade after *Desert Solitaire,* reveal a tension between Abbey's ten-dency to find mystery in the desert and his disavowals of anything that can be called supernaturalism. In his account of a trip down the Rio Urique in the Sierra Madre he describes watching mysterious lights floating around him in the dusk before realizing that they are fireflies. The incident becomes an occasion for rejecting Castaneda's "magic-button spooks" and "occult visions" in general; "To hell with mysticism," he concludes (*AR,* 90).[20] Yet Abbey can talk about the "hot mystic stillness of the desert" in the essay that precedes this one in the collection (*AR,* 80), and he finds mystery in moments that would seem commonplace to others. Lying with friends in the sun on the rim of Escalante canyon, he imagines a powerful and unknowable presence dominating the landscape:

> Over the desert and the canyons, down there in the rocks, a huge vibration of light and stillness and solitude shapes itself into the form of hovering wings spread out across the sky from the world's rim to the world's end. Not God—the term seems insufficient—but something unnameable, and more beautiful, and far greater, and more terrible. (*AR,* 120)

This is something beyond the mystery of the obvious, a suggestion of a presence much larger and more powerful than anything one could expect to find in the visible scene. One could call it a kind of desert sublime, arousing awe and fear, which Abbey responds to despite his hard-boiled rejection of the supernatural and the fraud-ulently mystical. In his classic *The Desert* (1901), which Abbey admired, John Van Dyke asks, "What is it that draws us to the boundless and the fathomless?" and concludes that it is the sublime that we feel in "immensity and mystery."[21] Abbey does not invoke the sublime so readily, and his temperament was profoundly differ-ent from Van Dyke's, yet his response to the mysteriousness of the desert resembles Van Dyke's and may have been shaped by it. Van Dyke anticipated Abbey and other writers on the American desert in his emphasis on the desert's silences, the "splendor of its light," and the paradoxical grandeur of its desolation.

Abbey's transient vision of hovering wings holds a suggestion of terror as well as of beauty because the immense stillness conveys a sense of nothingness that seems wholly alien to ordinary human consciousness and points to the annihilation of consciousness in death. In an interview Abbey once said, "I am terrified and at the same time fascinated by solitude, silence, and death," and went on to comment on his attraction to the "silent tension between death and life" that he found in the desert.[22] One sees the danger in the appeal of the nothingness Abbey finds in the desert in his attraction to the abyss, particularly apparent in the Grandview Point chapter of *Desert Solitaire.* Searching along the rim of the canyon for the tourist who will eventually be found dead, he looks down into the "awful dizzying vacancy" and imagines that the man might have gone over the edge deliberately, "spellbound by that fulfillment of nothing-ness" (*DS,* 238). The passage suggests Abbey's own romantic attrac-tion to death, seen as the ultimate way of merging with the other world of the desert. His frequently stated wish to die on a rock in the desert, which he tried to realize in fact, is one form this attrac-tion takes. Here it takes the form of envy of the tourist's actual manner of dying: "alone, on rock under sun at the brink of the unknown, like a wolf, like a great bird . . . before this desert vast-ness opening like a window onto eternity" (*DS,* 240). Jeffers antici-pated Abbey in this wish, imagining himself crawling out on a ledge to die like a wolf ("The Deer Lay Down Their Bones"), as he had in the wish to become a part of the vulture he imagined feeding on him ("Vulture"). If Abbey yields to the appeal of mysticism here, it is a "hard" kind closer to that of Jeffers than to the hallucinatory sense of merging with the landscape that he exorcised in the Havasu chapter. One sees something of Abbey's telltale humor in his habit of identifying with vultures (he tells us that he likes them because they are ugly and indolent), but the humor masks another kind of attraction, apparent in the way he imagines the death of the tourist: "he may have died in his sleep, dreaming of the edge of things, of flight into space, of soaring" (*DS*, 240). The remarkable image with which he ends this chapter enacts a fantasy of viewing himself "sinking into the landscape, fixed in place like a stone" through the eyes of a soaring vulture whose progressively widening vision eventually extends to the margins of the earth and "that ultimate world of sun and stars whose bounds we cannot discover" (*DS*, 244). The fantasy allows Abbey to realize his urge to become a part of the desert world yet preserve an individual consciousness, characteristically drawn by the appeal of a world without apparent limits. The overall effect of the image is to diminish the importance of the human presence and to link the desert with an enduring world beyond our capacity to know.

One of Abbey's favorite poses is on the rim of a canyon, looking down at the river thousands of feet below, perhaps dangling his feet into space as he describes himself doing at Cape Solitude. In the Grandview Point chapter he discusses what is more often implicit in his work, the romance of the abyss. The nothingness of the desert—suggested by what Abbey would have seen as "the depths of another world" (*JH*, xiv)—here offers a dangerous promise of fulfillment. In a journal entry written in a period of idleness and depression (in 1968), Abbey describes a "sick, secret, furtive longing for . . . that mystic landscape—that ultimate loneliness—that final peace." He embraces his melancholy and imagines "not suicide but—to *vanish.* Disappear. Fade off into the wilderness, never return—that decadent romanticism of nihilism haunts me still" (*Conf.,* 213). The appeal of disappearing into the "mystic landscape," in the manner of Everett Ruess, resembles that of the abyss. It is the appeal of solitude and of a peace dependent upon disengaging from the familiar, limited, corruptible, and frequently depressing world to become a part of a seemingly pure and timeless one. Abbey pulls himself back with language that shows his self-consciousness about such impulses ("sick," "decadent") at the same time that he appears to yield to them. He did not really want to disappear, of course, or to look too long into the abyss. But he liked to live on the edge of that other world, to push himself to his physical and psychic boundaries.

I have not commented on Abbey's sense of the beauty of the desert, but this is part of the explanation of his passion. He could go so far as to assert, in "A Writer's Credo," that he wrote "to praise the divine beauty of the natural world" (OL, 178).[23] Abbey may have lacked Van Dyke's eye for nuances of color and form, but he was capable of precise, evocative descriptive writing and committed to rendering the surfaces and appearances that defined the desert's appeal for him: red dust and burnt cliffs, the look of a particular juniper, the "sweet cool clear green light of dawn" (DS, 234) before the sun blazes down. Abbey often invoked more general terms (grand, magnificent, splendid) to convey wonder in the presence of beauty, drawing on the vocabulary of the sublime. His vision of hovering wings spread out across the sky suggests awe before an unnameable presence "more beautiful" as well as "more terrible" than God (AR, 120), but the desert's beauty does not have to be majestic for Abbey. It is invariably associated with wildness, and it pervades the landscape: "Everything is lovely and wild, with a virginal sweetness" (DS, 11).[24]

Abbey's renderings of the beauty of desert scenes are evocative because of the emotion with which he invests them. For someone who cultivated an image of surliness Abbey shows a surprising fondness for the adjective "sweet," especially in the privacy of his journal. In an entry for 1957 he invokes the Navajo concept of walking in beauty and, after describing the "magnificence" of "red, naked rock" and mountains "blazing with snow," exclaims, "Sweet sweet wilderness!" (Conf., 140). Abbey's fear that development and exploitation would overwhelm what he saw as "this wild innocent and defenseless beauty" intensifies the feeling here and in many of his descriptive passages. He saw the desert as painfully vulnerable, even in a place as remote as the Pinacate. Abbey's sense of the beauty of desert places is nowhere more acute than in the account of his river trip through Glen Canyon just before it was flooded. Looking back at that trip in an essay written years later, he indulged his nostalgia for Glen Canyon as it was then, "the wilderness alive and sweet and charged with mystery, miracle, magic" (*AR*, 118). In *Desert Solitaire* Abbey describes one of the benefits of his job at Arches as "the discovery of something intimate . . . in the remote" (45). Many of Abbey's descriptive passages suggest an effort to establish an intimacy with the place, something more like a lover's relationship than that of observer to observed.[25] His surliness was directed toward those he wanted to protect the desert from, including his readers at times.

In describing the desert as "virginal" and "defenseless" Abbey was of course invoking a particular image of femininity. The desert is "lovely" and "sweet" because innocent and in need of protection. This is a "wild" innocence, paradoxically, as if offering something of the allure of Eve before the Fall. Abbey would not have recognized the kind of womanliness Mary Austin saw in her desert:

> If the desert were a woman, I know well what she would be: deep-breasted, broad in the hips, tawny, with tawny hair, great masses of it lying smooth along her perfect curves . . . [with] such a countenance as should make men serve without desiring her . . . and you could not move her, no, not if you had all the earth to give, so much as one tawny hair's-breadth beyond her own desires.[26]

Austin saw an idealized version of self-sufficient womanhood, perhaps of herself in the desert, reversing the usual tropes that imply male domination of the wilderness.

The desert's beauty was inseparable from its mystery for Abbey. Descriptive passages often begin with a rendering of detail (the fragile, fantastic appearance of Delicate Arch) and shade into the marvelous: Delicate Arch reminds us that "*out there*" is "a different world, older and greater and deeper by far than ours" (*DS*, 42). In much of *Desert Solitaire* Abbey seems preoccupied with this other world. He perceives it in the stillness he feels in the Maze, in the timelessness he senses in the landscape he looks out over from the canyon rim above Rainbow Arch ("like a section of eternity"). He seems to enter it, as nearly as one can, as he drifts through the "fantastic" and "visionary" world of Glen Canyon "in a kind of waking dream" (*DS,* 187). Abbey describes other river trips as dreamlike as well, but in this journey in particular, shadowed by the prospect of loss, he gives the sense of entering an ancient, pure, and sacred place. He describes himself at one point as floating "in effortless peace deeper into Eden" (*DS,* 183), suggesting by the allusion that he has entered a world prior to and undisturbed by human history. In one of his recollections of this trip Abbey remembers "passing through a world so beautiful it seemed and had to be—eternal": "Such perfection of being we thought—these glens of sandstone, these winding corridors of mystery, leading each to its solitary revelation—could not possibly be changed" (*DR,* 231). But, as he learned, everything changes, "and nothing is more vulnerable than the beautiful" (*DR,* 231). In memory, Glen Canyon becomes a dream of perfection and a symbol of the world before the Fall.

Abbey's references to Eden in this episode and elsewhere in his writing suggest a powerful desire to recover a past world, one that appeals because he sees it as pristine and also as "beyond us and without limit." He recognizes the nostalgia and romanticism in this kind of yearning for a perfect wilderness but would not dismiss it on this account: "The romantic view, while not the whole truth, is a necessary part of the truth" (*DS,* 190). Such yearning frequently colors writing about wilderness in America and can easily become sentimentalism. Abbey avoids this for the most part by signaling his self-consciousness about his own romantic yearnings and by continually returning to the shapes and textures of the "dogmatically real earth" to which he anchors himself. And by qualifying his romanticism with antiromantic gestures such as insisting that his "paradise" contains not only scorpions and sandstorms and cactus but "disease and death and the rotting of flesh" (*DS,* 190). He avoids one extreme by going to another, here by insisting on the toughness of this world and proceeding to ridicule traditional Christian images of a changeless, perfectly ordered heaven. His outrageousness makes space for expressions of wonder and of longing for the kind of permanence and timelessness he perceives in the natural economy of the desert (if not in its individual life forms), while rejecting the permanence promised by a supernatural order.

In the same journal entry in which he discusses the "decadent romanticism of nihilism" Abbey characterizes his yearning in an unguarded and particularly revealing way: "Melancholia—dread—the sick sad heartsick yearning for something lost, remote, past, future, forever out of reach. An absolute love for something absolute. Infinitely good and sweet and pure and beautiful—immutable. Forest—mountain—desert—sea!" (*Conf.*, 214). As usual, Abbey labels such yearning "sick," but he nevertheless indulges it here, as if the need is all the more intense for being attached to an ideal that is "forever out of reach." It was important to him to preserve the ideal of absolute, immutable wilderness—as it was for Keats to see the image of a perfect, unattainable love in the figures on a Grecian urn. While the romance of wilderness may have been particularly strong at this period of Abbey's life (the journal entry is from January 1968, shortly following the publication of *Desert Solitaire*), the passion apparent here continued to surface in his writing in various forms. One of its most interesting manifestations, and the final one I will consider, can be found in Abbey's apocalyptic visions of the decline of cities and other human works and the resurgence of wilderness.

Abbey was attracted to ruins: ghost towns, abandoned cabins or miners' camps, and other evidence of human presence. Such traces, like the petroglyphs and the remains of cliff dwellings left by earlier inhabitants, were reminders that the desert held its stories. They were also emblems of the transience of human presence in the desert. Abbey's futuristic novel *Good News* (1980) plays out a fantasy of the collapse of urban civilization, with horrific images of life in the ruins of a Phoenix where the only order is provided by a sadistic military dictator. Much of the power of the novel comes from the dislocation of the familiar: dangling signs with messages that have become ironic, abandoned cars, disintegrating stores and office buildings. Abbey's ultimate revenge on the economic and technological society that he tried in so many ways to subvert was to imagine it as a literal wasteland: "Ruins. Ruins. All in ruins. Coyotes slink among the blackened walls, hunting rats. Anthills rise, Soleri-like, from the arid fountains of the covered mall. Young paloverde trees . . . grow from cracks in the asphalt of the endless parking lots."[27] If Jack Burns and the others who resist the military tyranny never make it to the refuges they envision in the countryside, the natural world nevertheless remains the locus of hope and value in the novel. Abbey imagines the collapsing city surrounded by "the murmuring stillness of the desert, the eternal dialogue of the wind and the sands" (*Good News,* 60), with mountain ranges beyond suggesting an enduring landscape that magnifies our sense of the impermanence of the urban society that we see unraveling.

In a late essay, "Theory of Anarchy," Abbey prophesies the collapse of the present structure of society "within a century" and the emergence of scattered human populations living naturally in the wilderness by hunting, gathering, and small-scale farming and meeting yearly "in the ruins of abandoned cities for great festivals of moral, spiritual, artistic, and intellectual renewal" (*OL,* 28).[28] This wishful vision, adumbrated in the prologue of *Good News,* suggests that the only way cities can serve the renewal of human community is by providing good ruins. The implication is that the surrounding wilderness will reassert its primacy and the surviving people reaffirm their relationship to it in their festivals. Visions of ruined dams complement those of ruined cities. When Abbey is not suggesting ways of blowing up Glen Canyon dam, he is predicting its natural demise in two hundred years through the silting up of Lake Powell. Then this paramount symbol of human interference with natural processes, "its penstocks blocked, will be transformed into a splendid waterfall. Good news!" (*OL,* 89). The gospel according to Abbey is an ironic one, promising that the engineers and developers and politicians will be defeated in the end by natural forces that they only imagined they controlled.

Abbey's most suggestive and powerful visions of the decline of civilization and the durability of wilderness are the early ones, in which he gave his romantic imagination more play. The best example is the stunning scene at the end of the "Water" chapter of *Desert Solitaire,* in which, after invoking the obsession of the "Developers" with what they perceive as a "desperate water shortage," he imagines time and the winds burying "Phoenix, Tucson, Albuquerque, all of them, under dunes of glowing sand, over which blue-eyed Navajo bedouin will herd their sheep and horses" (*DS*, 145).[29] In the canyons of Utah they will see "great waterfalls plunge over salt-filled, ancient, mysterious dams" (*DS*, 145). The irony here comes from submerging human works in scenes of the sort that Abbey celebrates for their beauty and mystery elsewhere in *Desert Solitaire*. Dams recede into the landscape like the cliff dwellings of the Anasazi, symbols of an unrecoverable human past. This kind of irony disappears in the final paragraph, in which Abbey's focus shifts to the natural sources of water in the desert and, in a haunting final image, to washes "where the community of the quiet deer walk at evening up glens of sandstone through tamarisk and sage toward the hidden springs of sweet, cool, still, clear, unfailing water" (*DS*, 146). The image makes Abbey's point that the desert has water sufficient for *its* needs, if not for ours, but it also suggests in a particularly lyrical way the peace and beauty and innocence that he associates with certain desert places, like the side canyons of the Escalante. The magic here depends upon erasing signs of human presence altogether. The springs are "hidden" from *us;* we have to imagine them. The desert must have its secrets and its private places, if it is to embody the kind of mystery Abbey wants to find there.

The most radical of Abbey's apocalyptic visions appears in "Dust: A Movie," the expressionistic sketch that concludes *The Journey Home.* Here we see the gradual deterioration of a desert ghost town through the camera's eye, which lingers on details that suggest the progressive intermingling of natural and human: "prickly pear growing from an earthen rooftop" (*JH*, 241), aimless human footprints among tracks of coyote and fox. The primary actor in this imaginary movie is the natural world, with destructive winds and waters (including a flash flood) that assault the remains of buildings and are responsible for the death of the couple that we see wandering in the ruins. This is the closest Abbey comes to the kind of misanthropic vision of a landscape cleansed of human contamination that one finds in Jeffers (e.g., "November Surf"). Through a series of dissolves he shows the town reduced to earth mounds and sunken stone walls and then disappearing into the desert. Bleached human bones are washed away. Again Abbey ends with deer, here coming to drink in the dawn, watched by a mountain lion. The final image of the lion looking into the camera, which then zooms in so that its eyes fill the screen, becomes a romantic celebration of the desert world: "We see in their golden depths the reflection of the sunrise, the soaring birds, the cliffs, the clouds, the sky, the earth, the human mind, the world beyond this world we love and hardly know at all. . . . DISSOLVE. This film goes on, it has no end. . . . DISSOLVE . . . DISSOLVE . . . DISSOLVE. . ." (*JH,* 242). By this seemingly infinite regression, Abbey suggests not only that the natural world mirrored in the lion's eyes has no end but that we are absorbed into it. The camera's eye itself seems to dissolve.

One comes away from Abbey's work with a sense of the powerful presence and the enduring appeal of the desert, what he calls in *Desert Solitaire* the "golden lure" that makes one "a prospector for life" (272). For all of his antiromantic gesturing and his charting of the failures of vision, Abbey never cured himself of his romanticism. It is unlikely that he really wanted to, since it is the basis of a view of the natural world and our place in it that is finally optimistic, despite his consciousness of loss and of a continuing pattern of destructive human intervention. Abbey characteristically undercuts the drama of the scene in which he takes his leave of Arches, relinquishing "possession," by insisting upon the indifference of the desert landscape to his own or any other human presence. This recognition, however, does not prevent him from offering the parting gesture of optimistically imagining the landscape reviving after the destruction of nuclear holocaust: "The canyons and hills, the springs and rocks will still be here, the sunlight will filter through, water will form and warmth shall be upon the land and after sufficient time, no matter how long, somewhere, living things will emerge and join and stand once again" (*DS,* 301). Joining two of his favorite terms a final time, Abbey describes the "animal faith" that makes this vision possible as "bedrock," "close by the old road that leads eventually out of the valley of paradox" (*DS,* 301). Such faith in the continuity of the land and the life it nurtures becomes a way of countering the resistance of the desert to rational understanding.

If the desert remains a riddle for Abbey, he can at least believe in its vitality and endurance.

Scenes showing the continuity of wild nature and the impermanence of human works express Abbey's bedrock optimism. Toward the end of *Desert Solitaire* he says that "life nowhere appears so brave, so bright, so full of oracle and miracle as in the desert" (286). If he was drawn to the edge of the abyss and to the sense of nothingness he found in the vast silences of the desert, he also responded to such evidence of life as the noisy revival of spadefoot toads after rain (*DS*, 142–44). Abbey could be most optimistic when he edited out signs of human presence altogether, as in his rendering of the "choral celebration" of the toads and his evocation of "the community of the quiet deer." We respond to the image of the spadefoot toads because of the way his imagination colors the scene, as he violates his injunctions against personification to show them singing "out of spontaneous love and joy . . . for love of their own existence, however brief it may be, and for joy in the common life" (*DS*, 143). Although Abbey argues that such joy has survival value, implying that he has some kind of rational justification for attributing it to the spadefoot toads, it is at bottom an expression of his will to believe in the resilience of the desert. And of an inveterate romanticism, apparent also in his subsequent celebration of the natural world's capacity for renewal ("The rains will come, the potholes shall be filled. Again. And again. And again") (*DS*, 144). Abbey thought that a writer should be driven by passion and that this should be "fueled in equal parts by anger and love" (*OL*, 176). His love for wilderness found expression in a romanticism that he often deprecated or restrained but that is as fundamental to his vision of the world as the anger that found an outlet in devastating satire.

5 Into the Woods with Wendell Berry

I go in pilgrimage
Across an old fenced boundary
To wildness without age
Where, in their long dominion,
The trees have been left free.

——WENDELL BERRY,
A Timbered Choir

Wendell Berry commands attention as a passionate and eloquent defender of sustainable agriculture on a human scale, a morally as well as economically viable farming that implies respect for the land, for family and community, and for the wisdom embodied in local culture. Through his fiction, his poetry (including *Farming: A Handbook,* 1970), and especially collections of essays such as *The Unsettling of America* (1977) and *The Gift of Good Land* (1981), Berry has become widely known as a persuasive defender of life and work rooted in a particular, rural place (in his case a hill farm in Henry County, Kentucky) and as a trenchant critic of agribusiness and of what he would call the industrial as opposed to the natural economy. Yet one of the most striking things about Berry's work is his attraction to wilderness, particularly in the form of the forests from which the cleared fields of the farmer were wrested and to which he imagines them eventually returning.

Berry resists the common tendency to oppose nature and culture, the wild and the domestic, and finds meaning and health in their interaction. In his essays, beginning at least with *A Continuous*

Harmony (1970), Berry stresses the interdependency of nature and culture and urges a life based upon achieving a harmony between these apparent opposites. In "The Body and the Earth," published in *The Unsettling of America,* he faults modern travelers for transforming wilderness into "scenery" and forgetting that it "still circumscribe[s] civilization and persist[s] in domesticity."[1] Such a claim reverses the familiar perspective that assumes human dominion over the natural world to suggest that wilderness contains civilization and continues to be felt within the space claimed by it; Berry implies that we should recognize that we are still part of a dominant natural order.

Berry's insistence that the civilized and the domestic depend upon wilderness reflects deeply held beliefs about both the health of the farm and the health of the spirit. If the farm is to last and to thrive, the wilderness must survive within it.[2] In the most practical sense, this means recognizing that fertility is linked to the natural processes of growth and decay that the wilderness exemplifies. One must look to the woods to know how to preserve the fields, an insight Berry attributes to the English agriculturalist Sir Albert Howard but has made thoroughly his own. Berry offers practical prescriptions for a "good forest economy," based upon sustainable, small-scale logging. At the same time he urges that substantial tracts of wilderness be preserved, as a standard: "Wilderness gives us the indispensable pattern and measure of sustainability."[3] Berry also urges the importance of preserving a part of the actual forest within the farm as a "sacred grove," "a place where the Creation is let alone, to serve as instruction, example, refuge."[4] Preserving such wild places within the human order assumes great importance for Berry because he regards them as places of meditation and of spiritual solace and renewal. They function as continual reminders of the mysteriousness of the natural order and the limitations of human understanding and capability.

In some of his essays, particularly "Getting Along with Nature" and "Preserving Wildness" in *Home Economics* (1987), Berry insists even more strongly on what he calls "the indivisibility of wildness and domesticity," partly in an effort to stake out a middle ground between the "nature extremists," who assume that natural good is human good, and the "technology extremists," who would manipulate nature to serve their particular sense of human good.[5] Berry can accept the argument for setting aside public places of "absolute wilderness" (Abbey's term) but is more interested in the small pieces of wilderness found on farms or in corners of cities that he views as sacred groves "where one can go to learn and be restored" (*HE*, 17). Even margins, such as fencerows or streamsides, are valuable "freeholds of wilderness" (*HE*, 151) that help to create a "landscape of harmony," preferable to the "landscape of monoculture," in which the wild and the domesticated coexist. One finds a growing concern about the survival of wilderness in Berry's later writing (expressed in "Preserving Wildness"), a concern that wilderness has become dependent upon human forbearance, but this only intensifies his conviction that one must find ways of valuing and preserving the wild in daily, domestic life. As what Berry calls "our native wilderness" comes to seem more vulnerable, his advocacy of living in harmony with it has become more urgent: "I am betting my life that such a harmony is possible" (*HE*, 138).

What interests me here is the persistent and vivid presence of wilderness in Berry's work, not as an alternative to or escape from ordinary life but as a source of illumination, order, peace, and joy that helps one to understand and sustain and finally to leave this life. Berry invites his reader to go into the woods with him over and over in his work, presenting this act variously as journey, pilgrimage, meditation, and worship. My concern is with what the woods (and, more broadly, the wilderness) come to mean for him and why engagement with them assumes such fundamental importance in his writing. What does it mean to go into the woods with Berry? How does he reconcile, or attempt to reconcile, claims of wildness and of domesticity that many have found in conflict? How can he embrace the values he associates with agrarian life and still yield in so many ways to the attraction of wild nature?

Wilderness can take many forms for Berry, including an empty lot in which a few locust trees attract warblers and tanagers, "woods birds . . . wild as leaves" ("The Wild," *CP*, 19–20), and the "teeming wilderness in the topsoil" (*HE*, 11).[6] It can become a metaphoric way of referring to elements of human nature, as in

"The Body and the Earth," where one form of wilderness is the instinctual sexuality that Berry sees as a vital component of marriage. In the early autobiographical essays Berry collected as *The Long-Legged House* (1969) wilderness can mean the Kentucky River beside which he writes, "the power and mystery of it" as it rises in flood (*LLH,* 10); the natural setting of the Camp, the family "wilderness place" that he knew from childhood and to which he returned with his wife Tanya the summer after their marriage; and, especially, the woods of the "native hill" to which he climbs from the fields he farms.[7] In these essays and in his early poetry Berry began to articulate a sense of the nature of wilderness and of the ways humans have dealt with it and can find their place in relation to it, introducing themes that would recur and deepen throughout his work.

One of the most distinctive aspects of Berry's wilderness is its historical dimension. For him the land is haunted by "the ghost of the old forest / that stood here when we came" ("Planting Trees," *CP,* 155) and that has largely vanished through acts of exploitation and negligence. The "we" here represents the white settlers and their vanguard the long hunters, Daniel Boone among them, who came over the mountains from Virginia lured by the prospect of abundant game and fertile land, and their by-now-numerous descendants, including Berry himself. Berry can imagine the wonder of discovery, when the long hunter emerging from his "unknown track" first sees the new country opening out beneath him: "from the height / he saw a place green as welcome / on whose still water the sky lay white" ("The Long Hunter," *CP,* 168). The early settlers of "July, 1773" (*CP,* 221) find at a crowded salt lick an "upwelling / and abounding" of herds so astonishing that it seems a "holy vision." Yet the scene turns to one of foreboding as young Sam Adams shoots into a herd of buffalo (he "could not let it be") and stampedes them, at great risk to himself and his uncle. The act serves as an image of violation (they have been warned to respect the hunting ground of the Shawanoes), implying the future cost of white settlement. Berry assumes that the land is pristine to this point, undamaged by the activities of native tribes. The wonder of the unspoiled land is inseparable for him from a consciousness of the violence that has been done to it and to its original inhabitants, what he sees as the "burden" of its history. The effort to reconnect himself to his native place and the history of his family in it that Berry describes in *The Long-Legged House* is painful; he recognizes the progress brought by settlement as dooming what he values most: "the life and health of the earth, the peacefulness of human communities and households" (*LLH,* 179). He deals with the pain of loss by working to heal land damaged by abusive farming and by finding a way of relating to the natural world with respect and humility. Berry's way will not be the destructive "way of the frontiersman" but the "way of Thoreau, who went to the natural places to become quiet in them, to learn from them, to be restored by them" (*LLH,* 42).[8] Ultimately, he salves the pain by imagining a wilderness so durable and encompassing that it will outlast human efforts to domesticate it.

To justify what he calls the "way of Thoreau" Berry must find in wild nature a mystery and natural goodness uncompromised by the occupation of white settlers bent on dominating it. When he hears pileated woodpeckers from a canoe, he imagines these birds of "big trees and big woods" as speaking "out of our past" (*LLH,* 107), evoking for him by their wild cries a time before the power-boating and strip-mining that he inveighs against in other parts of *The Long-Legged House.* He finds vestiges of this time in the tall open groves that he sees scattered on the slopes of the river valley, "as if in a profound interior withheld from the ceaseless drone of engines, a fragment of the great quietness that two centuries ago lay upon the whole valley" (*LLH,* 33). Berry must edit out the noise of engines, from traffic on the road as well as power boats on the river, to suggest the silence with which he invariably associates inviting natural places. Here these are groves of "solemn beauty" with "delicate flowers and mosses and ferns" on the woodland floor. In them one seems to enter a zone of quiet, which is as much a meditative as a physical state. The "voice" of these groves for Berry is the wood thrush, "whose notes, without replacing the quiet, flow into it from some hidden perch" (*LLH,* 33). If one responds to such charged "quiet" with a meditative quiet of one's own and sufficient attentiveness, the woods will reveal its creatures. What Berry describes here and elsewhere in *The Long-Legged House* is a state of awareness that he comes to recognize as a condition of the knowledge and restoration of the spirit that he associates with Thoreau.[9]

Such an emphasis upon respectful awareness is not uncommon among writers who focus on human relationships with the natural world. One can find it in different forms in the writing of Annie Dillard and Barry Lopez, among others. What is unusual is Berry's pained consciousness of the damaging effects of human presence and his effort to recover a vision of the natural order that preceded this damage, an undisturbed wilderness (the "old forest") that he associates with the time prior to white settlement. This can never be an untroubled vision for Berry, as his poem "The Dream" (*CP*, 64) suggests. There he describes a dream of taking away roads and fences and machines and restoring the great trees of the "first forest" ("I glimpse the country as it was") that turns bitter as he is unable to stop the "flowing in" of a human history of inevitable ruin. Berry can find fragments of the "great quietness" of the natural world of two hundred years ago, as his experience of the old groves in *The Long-Legged House* suggests, but his appreciation of these remnants is colored by deep nostalgia and a tragic sense of what has been lost. In one of his earliest published poems Berry pictures Daniel Boone in old age recognizing the elusiveness of the vision he pursued: "We held a country in our minds / which, unpossessed, allowed / the encroachment of our dreams" ("Boone," *CP*, 8). To possess the land is to destroy those qualities that made it so alluring in prospect.

Berry gives his wilderness a future as well as a past by imagining it as resilient and vigorous enough to heal itself and ultimately efface the signs of human occupation. He finds hope in the recuperative powers of the natural world that he has observed around him. If the effects of bad farming practices are apparent in eroded slopes and the silting and frequent flooding of the river, so are the failures of the farmers responsible for them to sustain their enterprise. Where the land has been abandoned, the woods return: "The wild is flowing back like a tide" ("A Native Hill," *LLH*, 188). Berry frequently suggests the fragility of the farmer's hold on his land, sometimes in images of the scrub growth that overruns neglected fields, the "snarls and veils / of honeysuckle, tangles / of grape and bittersweet" ("The Clearing," *CP*, 180). In the short run, Berry the farmer views his own clearing of such fields as restoring their health by making them productive again. He takes pleasure in sowing the seeds that will result in a rich pasture of clover and grass and identifies with the succession of those who farm the land with a sense of how it should be used. Yet when he takes a longer view, he can identify with the grandfather of "The Handing Down" (*CP*, 44) who knows that "there's a wildness / waiting for him to go" and that eventually "the woods take back the land" as the big trees return. In a striking late poem collected in *A Timbered Choir* (1998) ("I walk in openings / That when I'm dead will close" [119–20]), Berry describes himself stopping his mower blade to spare the seedling trees invading his meadow, by this gesture affirming the larger natural order to which he submits. This perspective, which accepts oak and ash and hickory as "the genius of this place," enables him to value clearing and farming the land without devaluing wilderness, as Jefferson and Crevecoeur devalued it in their visions of an agrarian America. Berry's confidence that the big trees will indeed return could scarcely be more unlike the fatalism with which Faulkner shows the Big Woods of *Go Down Moses* inexorably giving way to cotton fields.

Berry can accept the natural process by which the big trees return as the ultimate healing of the land because he views human works as impermanent; the inevitable return of the woods offers a lesson in humility. One measure of Berry's sensitivity to the history of interaction with the land is his fascination with the signs of former human activity: old house sites or stonework in streams, overgrown wagon tracks, flowers planted by some farm wife that persist after the farm itself is gone. In *The Long-Legged House* he imagines the country as "a kind of palimpsest scrawled over with the comings and goings of people, the erasure of time already in process even as the marks of passage are put down" (185). This powerful image of human transience seems particularly appropriate to Berry's rural Kentucky or to Frost's New England with its cellar holes and stone walls meandering through the woods, less apt to urban America. One would find it harder to believe in the kind of "erasure" Berry describes in Manhattan or Los Angeles. While he is capable of picturing the forest returning to "the waste places of the cities" and "wild herds" on the highways ("Window Poems," *CP,* 83), such apocalyptic imaginings seem more natural in Abbey's fantasy of cities buried by sand dunes or Walker Percy's extended fantasy of the collapse of urban civilization in *Love in the Ruins,* where one sees vines engulfing the A&P and breaking through the pavement of the interstate.

The autobiographical essays of *The Long-Legged House* are most effective as a record of how Berry comes to appreciate what it means for him to be a "placed" person, rather than the kind of displaced person he finds more typical of modern America. In these essays one discovers how to live "within rather than upon the life of the place," "to belong as the thrushes and the herons and the muskrats belonged, to be altogether at home here" (*LLH,* 150). This process grows out of Berry's earlier experience of returning periodically to enjoy the solitude and sense of freedom the place made possible; it is inseparable from the experience of restoring and then rebuilding the Camp and of settling into his marriage there. It involves, crucially, learning the "power of silence" and attentiveness and establishing the sense of trust, like that demanded by a marriage, that grows out of his commitment to make his life in the place. Then, with familiarity and trust and attentiveness, "the mind goes free of abstractions, and renews itself in the presence of creation" (*LLH,* 160). The kind of seeing that this state makes possible has the character of a sudden revelation granted an initiate: "the details rise up out of the whole and become visible" (*LLH,* 160), as a hawk stoops into a clearing and a wood drake swims out of his hiding place. In Berry's writing such experiences inevitably take on a religious coloration. He speaks of his "devotion" to the place, of his "holy days," of being born there in spirit as well as in body.

Berry recognizes that the kinds of questions he asks about the nature of the place and how he should live in it are religious because they acknowledge the limits of his understanding, at the same time distancing himself from the tendency of the organized religion he knows to thrive on what he regards as "a destructive schism between body and soul, Heaven and earth" (*LLH,* 199). Such questions are for him "an enactment of humility" (*LLH,* 199), acknowledging and honoring the mystery he finds in the created world. To ask them and to use religious language in describing his experience is a way of rejecting and transcending the assumptions of orthodox religion, as his insistence that he belongs to the place is a way of rejecting or at least displacing the language of property ownership and the history of violence that he sees possession of the land as having entailed. What Berry offers in describing his most intimate and profound experiences of the natural world is a distinctive kind of spirituality. Its implications emerge most fully from the meditation on the experience of entering the woods that he describes near the end of "A Native Hill."

Berry represents himself as leaving the landmarks of civilization (barn, old roads, cow paths) to enter the "timelessness" of a wild place, here woods with large trunks rising free of undergrowth, not the great beech trees of the old forest but a reminder of them. They offer qualities Berry characteristically associates with woods: dignity, serenity, order, joy. He describes the delight of finding the floor of these woods covered with bluebells one spring morning. The image, with the sense of "joyful surprise" associated with the discovery, prints itself on the memory like Wordsworth's daffodils and serves as a reminder that the world is "blessed" beyond his understanding. Berry's delight in spring woodland flowers, with their delicate beauty and their proof of the annual renewal of life, recurs throughout his work as a sign of the blessedness of the natural world. In a poem collected in *A Timbered Choir,* the "saving loveliness" of "bloodroot, / Twinleaf, and rue anemone" offers an assurance of permanence despite "human wrong."[10] As he enters the woods—or "re-enter[s]," as he puts it—Berry imagines himself as moving out of the space and time of human domination of the natural world, out of the conditions of modern life, into a prior state: "One has come into the presence of mystery . . . one has come back under the spell of a primitive awe, wordless and humble" (*LLH,* 205). Berry describes himself at one point as a "pilgrim" coming to a "holy place." His primitivism is qualified by the sense that the presence he encounters is "the presence of creation," where he finds the "perfect interfusion of life and design" (*LLH,* 201). Berry can enter the woods so humbly and reverentially, in other words, because he finds divinity in the order and peacefulness and beauty he discovers there.

His way of representing his experience of the woods recalls the negative way of Christian mysticism: "only there can a man encounter the silence and darkness of his own absence" (*LLH,* 207). Henry Vaughan's "The Night" describes the soul retreating from the busyness of the "loud" daytime world to discover God's "deep, but dazzling darkness."[11] The silence and darkness that Berry associates with the experience of the woods, here and repeatedly in his work, suggest the necessity of emptying the mind and disengaging oneself from the preoccupations of ordinary, daily life as preparation for this experience. Yet for Berry this process leads not to the mystic's sense of transcendence of this world and spiritual union with God but to a greatly heightened awareness of natural phenomena and a sense of harmony with the world to which they belong: "As its sounds come into his hearing, and its lights and colors come into his vision, and its odors come into his nostrils, then he may come into *its* presence as he never has before, and he will arrive in his place and want to remain. His life will grow out of the ground like the other lives of the place, and take its place among them" (*LLH,* 207). One is transformed by entering the silence and darkness of the woods, "born again," as Berry puts it. At this stage of his career, such a rebirth means to understand and embrace the condition of being rooted in one's native place and, as we have seen, to develop the capacity to live "within" it. To live in the sense of immersing oneself fully in the present moment and experiencing a heightened sensuous awareness. The promise of such a life is for Berry something like the "great peacefulness" he sees in the lives of wild creatures, wood ducks dozing in the sun or a flock of cedar waxwings eating wild grapes: "though they have no Sundays, their days are full of sabbaths" (*LLH,* 211). The term, which becomes richly symbolic in Berry's later poetry, embodies here a yearning to enter a state of rest and freedom from anxious thought authorized by the natural order of creation.

Berry's most sustained meditation on wilderness can be found in the collection of essays on Kentucky's Red River Gorge that he published with photographs by his friend Gene Meatyard in 1971 and reissued in 1991 (*The Unforeseen Wilderness*).[12] This remarkable and insufficiently known work transcends its occasion, the eventually successful struggle to prevent the damming of the river. Berry describes the process of discovering the Red River country, through backpacking and canoeing trips over a period of several years, as "one of the landmarks of my life" (*UW*, 60).[13] The essays register his growing appreciation of the place as he passes from his initial unease and sense of "strangeness" to a familiarity that makes possible the feeling of surrendering to a place grown friendly in the solitary walk with which the book ends. Berry's sense of coming to find his own place in the landscape recalls the process he describes in *The Long-Legged House,* but this experience differs in that it involves the "ancient fear of the unknown" and the "essential loneliness" that he associates with going alone into "big woods" (*UW*, 29). The place itself, although marked by traces of human use in most of its parts, comes closer to what Berry calls "the creation in its pure state," his most fundamental definition of wilderness.

By giving a detailed record of his own observations and reactions and interweaving these with meditations on others he regards as having treated the land well (the original Indian occupants, Daniel Boone and the long hunters, a family presently farming in the gorge), Berry establishes a pattern of respectful use against which he can judge those who have abused the land in the past (miners and farmers out for quick profit) or who would abuse it in the future (the Corps of Engineers and developers selling the promise of recreation). He begins, in "A Country of Edges," by tracing his initial descent into the gorge in March, through an abundance of spring wildflowers and the omnipresent sound of falling water, conveying a sense of the overwhelming freshness of the landscape and of the power and "residual mystery" (*UW*, 6) of the always changing river as it continues to sculpt its banks. As with the woods of his native hill, Berry establishes a sense of humility before a natural order that offers more than one can comprehend "in one human lifetime"(*UW*, 6). In the account of a June canoe trip on the river with which he ends the essay, he dislocates the reader's normal time sense, moving from a description of idling on the river and becoming intimately acquainted with its features to a meditation on the huge boulders that have fallen into the river and in their "silence and endurance" (*UW*, 9) prompt thoughts of the generations (Indian and white) that have paddled around them. This leisurely introduction, establishing a sense of the presence of the place and setting out the beginnings of his own journey of discovery, creates a context for Berry's attack on the Corps of Engineers and other past and potential exploiters of the Red River country in his second essay, "The One-Inch Journey."

By returning to the legend of John Swift wandering in the Kentucky mountains in the mid-eighteenth century in search of silver, Berry links those who would dam the river and open the area to development with a history of abuse that includes mining, logging, and destructive farming practices. This is a history driven by the expectation of quick profit and oblivious to what Berry sees as the real life of the place, with its complex cycles of growth and decay. An unnamed family returned from the city to an ancestral farm in the bottomlands of the gorge provides Berry with a countervailing example of how to live on the land. He offers a romantic vision of the isolated farmhouse on the edge of the woods at the foot of a great cliff: "the quiet of the wilderness rises over it like the loving gray cliff face. . . . [The house] seems somehow to have assumed the musing inwardness of the stone that towers over it" (*UW*, 25). We learn nothing of the house itself except that it "has over the years been cleansed of all unnecessary sounds," as if moving into the silence of the wild country that surrounds it. To create such a vision Berry must erase the normal signs of human occupancy of the land, blurring the boundary between natural and domesticated spaces. The potential flooding of this bottomland, with the consequent destruction of the idealized family's house and fields, becomes an emblem of the public logic of a kind of development that will destroy its natural purity.

In a subsequent essay Berry offers another kind of romantic vision that complements this one, prompted by seeing a crude hut with "D. boon" carved on a plank that has become a spectacle for tourists and is now surrounded by a chain-link fence and littered with trash. Upon entering the one grove of virgin forest left in the

Red River country, Berry offers his own reconstruction of a hut
Boone might have had in a cave overlooking such a place, in a time
of "vast rich unspoiled distances quietly peopled by scattered Indian
tribes," as well as by "buffalo and bear and panther and wolf":

> Imagine a man dressed in skins coming silently down off
> the ridge and along the cliff face into the shelter of the
> rock house. Imagine his silence that is unbroken as he
> enters, crawling, a small hut that is only a negligible
> detail among the stone rubble of the cave floor, as unob-
> trusive there as the nest of an animal or bird, and as he
> livens the banked embers of a fire on the stone hearth,
> adding wood, and holds out his chilled hands before the
> blaze. Imagine him roasting his supper meat on a stick
> over a fire while the night falls and the darkness and the
> wind enclose the hollow. Imagine him sitting on there,
> miles and months from words, staring into the fire, let-
> ting its warmth deepen in him until finally he sleeps.
> Imagine his sleep. (*UW*, 71–72)

This remarkable vision embodies a yearning for a natural world "in
its pure state," not a world without humans but a world of primitive
simplicity that admits no sense of a division between man and
nature, a world before any human destruction. Berry virtually
effaces the hut, making it "negligible," "unobtrusive" as the dwelling
of a creature. The withdrawal from words, along with the thought
processes and human interactions language implies, suggests the
recovery of a simpler state of consciousness that makes possible the
perfect harmony of human figure and setting that the scene suggests. One could imagine other Boones: the colonizer who led settlers through the Cumberland Gap and founded Boonsborough, the
legendary Indian fighter, the land speculator.[14] But what clearly
matters to Berry here is the elemental character of the scene, with
the solitary hunter absorbed by the silence and vastness of the primal wilderness rather than disrupting it, as John Swift and his suc-
cessors would.

Berry opposes another kind of figure to those he accuses of

exploitation, one who typifies for him the kind of seeing that leads to understanding and appreciation of the land, the "photographic artist." The immediate contrast in this case is with the tourist photographer interested only in preconceived, "postcard" pictures. Berry's rendering of the "quest" of the true photographer obviously draws upon his exposure to Gene Meatyard's methods and his black and white photographs of the gorge. Just as obviously, it reflects his own sense of how he approaches the natural world as a writer. The photographer must have the humility to efface himself and must be open to the unexpected in order to "draw deeper into the presence, and into the mystery, of what is underfoot and overhead and all around" (*UW*, 29). He enters the darkness of the shadowy woods, and of his own uncertainty, in order to see: "Among the dark trees . . . there appears suddenly the tree of light" (*UW*, 28). Such moments have the character of revelations granted by the place, manifestations of grace. Berry comments in the foreword to the 1991 edition that the darkness in some of Meatyard's pictures, all of which involve the play of light and shadow, is not simply shadow but "the darkness of elemental mystery, the original condition in which we can see the light" (xii). A belief in the mysteriousness of the natural world, best symbolized by such darkness, is fundamental to Berry's sense of the spiritual character of the journey of the photographer and of anyone who would discover this world. This is not a journey to be measured by our usual notions of space and time, Berry suggests, but "a journey of one inch, very arduous and humbling and joyful, by which we arrive at the ground at our feet, and learn to be at home" (*UW*, 30).

In the central essay in the collection, "An Entrance to the Woods," Berry presents a solo trip into the gorge as a paradigmatic encounter with wilderness. The process of entry begins with an acute sense of dislocation as Berry absorbs the shock of his sudden transition from freeway speed to foot speed and from his usual places and human relationships to the solitude of a campsite by a creek at the bottom of the gorge. The emotional fallout of this dislocation is melancholy, the result of cutting himself off from the familiar and stripping away his "human facade" to enter the wilderness "naked," with no more than he can carry. Berry stresses the elemental nature of the experience, seeing himself as reduced to the "loneliness and humbleness" of unaccommodated man, and gives it the character of a religious quest. He must lose his normal identity ("I will be absorbed into the being of this place, invisible as a squirrel in his nest" [*UW*, 36]) and undergo "a kind of death" (*UW*, 38) in order to experience the spiritual rebirth that he seeks.

To enter the wilderness with Berry is to experience a sense of "nonhuman time" that makes human history seem insubstantial and to be haunted by the ephemeral presences of those who have preceded him in the place: Indians, long hunters, loggers, and farmers who have left only chimney stones and flowers marking a vanished dooryard. Berry's sensitivity to such ghosts is a further source of melancholy, but he appears to need to identify with them, particularly with the Indians and long hunters of the time prior to white settlement. To describe his tent as a "flimsy shelter" and to picture himself as like a squirrel in his nest suggest a desire to recover something resembling the condition he attributes to Boone in his imagined hut, the solitary human figure blending into the vast surrounding wilderness. To perform this act of recovery, Berry must overcome or at least limit his awareness of the civilization he has temporarily left, symbolized by the highway noise audible from most parts of the gorge. He does this in part by imagining wilderness as the enveloping element in which humans live, "encased in civilization," an inescapable primal nature that is "beautiful, dangerous, abundant, oblivious of us, mysterious, never to be conquered or controlled or second-guessed, or known more than a little" (*UW*, 37). Such a view attributes extraordinary power and resilience as well as romance to wild nature and minimizes the effects of human activity on it, more apparent today than when Berry wrote. He imagines an airplane, its engines or instruments failing, entering the sphere of the wilderness "where nothing is foreseeable" and "the power and the knowledge of men count for nothing" (*UW*, 36). This image of the fallibility of human technology is part of a broader attack on the machinery of industrial society that runs through Berry's writing. It focuses on the vulnerability of the machine—airplanes do crash in remote areas—rather than on its use to extend human power by minimizing the effect of distance and making the remote accessible. Berry must diminish the sense of a controlling human presence to sustain the view of wilderness he offers here. He recognizes that the gorge is in fact an "island of wilderness," and a threatened one at that, but he would make it an emblem of a fundamental reality that should serve as a check on human pretension.

In the narrative he constructs of his day of hiking in the gorge, Berry shows himself becoming progressively more attuned to the place, to its quietness and the movements of its creatures, and losing his sense of difference from it. When he climbs to a high, sunny rock to get out of the morning cold of the hollow, he delights in the sensations and in the freedom of being "afoot in the woods," although the roar of the highway as he looks out over the country triggers a meditation on the way engine noise has displaced natural silence on the continent; this modulates into a denunciation of the destructiveness of the American economy with an allusion to the war in Vietnam that has the effect of reminding the reader of the preoccupations Berry is trying to escape by coming to this wilderness place. The essay recovers its momentum as Berry shows himself regaining his calm and a measure of hopefulness by recalling his absence from human society and resuming his walk. He follows a more promising direction back into the woods, then off the trail and down a small branch of the creek, eventually picking his way from rock to rock "through a green tunnel" of overhanging vegetation. It is in this departure from the predictability of the trail, while sitting still to watch fish in a clear pool enveloped by a "grand deep autumn quiet," that Berry has the sense of having "come into the heart of the woods" (*UW*, 44). The process he describes is one of joining his own quiet of mind with the quiet of his surroundings. When he returns to his camp, his sense of union with the place is so complete that the sounds of the creek move through his mind "as they move through the valley, unimpeded and clear" (*UW*, 44). Achieving this sense of union, and the peacefulness it affords, requires slowing the body and emptying the mind and stripping away all vestiges of the desire to control and exploit that he associates with the history of white settlement. Berry defines his role as if in opposition to that of his self-assertive predecessors, effacing himself in acts of submission to show an alternative to the way of dominion: "I move in the landscape as one of its details" (*UW*, 42).

The remaining two essays, "The Unforeseen Wilderness" and "The Journey's End," describe a deepening sense of harmony with the place and of the kinds of revelations this offers. Berry pictures himself as a "student" of wilderness, alert to its continuing lessons: that it changes continuously, as much process as place; that in such a place one is wiser to proceed without particular expectations; that here one is continously being surprised, by a "rare flame azalea in bloom" revealed by a turn in the path or a little dell where several streams meet under great hemlocks and beeches (*UW*, 49). Berry takes such chance pleasures as blessings, evidence of an ongoing, divine work of creation that he increasingly opposes to what he sees as the human work of destruction. Berry acknowledges the religious turn of his thought in the final essay, recognizing that he is "celebrating the morning of the seventh day" and finding the creation good: "This is a great Work, this is a great Work" (*UW*, 71). He moves here from the sense of quiet and harmony he described earlier to a reverential joyousness in the "intelligence" of natural processes set in motion at the beginning of the world and continuing without regard for human action or understanding.

Berry ends *The Unforeseen Wilderness* with an acccount of a solo hike in December in an unfamiliar part of the gorge in which he sees himself as abandoning any sense of control or direction: "Since I have no destination that I know, where I am going is always where I am. When I come to good resting places, I rest. I rest whether I am tired or not because the places are good. Each one is an arrival. I am where I have been going" (*UW*, 73). He experiences arrivals of a kind that Boone, chasing his dream of a promised land, never could ("There are no arrivals"). The journey can end because Berry recognizes that the destination does not matter, only the ease he feels wherever he chooses to pause and the process of yielding before "the mysteries of growth and renewal and change" (*UW*, 74), a yielding symbolized for him by the actual loss of his map among the fallen leaves of the gorge. The image of the map rotting along with the leaves becomes the vehicle of an epiphany for Berry, a "cleansing vision" of the tragic inadequacy of human knowledge. He has reached the point where he can be guided by his responses to what he understands as a divine order embodied in the natural world, rather than by the design of the mapmaker. Berry leaves the reader with a sense that the dominant presence is the earth, however humankind may abuse it, the creation that "bears triumphantly on from the fulfillment of catastrophe to the fulfillment of hepatica blossoms" (*UW*, 74). His optimistic belief in the resilience of the earth, imaged here by the annual reappearance of one of the most delicate of spring woodland flowers, overcomes the sense of ruin and impending loss that has threatened the peace and reassurance he has sought in his solitary excursion.

The transformation that Berry describes in *The Unforeseen Wilderness* depends upon a radical separation from habitual physical and social environments that is unusual for him, but it is continuous with the process of recovery that he represents in the essays of *The Long-Legged House* and embodies attitudes toward wilderness that he elaborates in his poetry and, to a lesser extent, in his fiction. The peacefulness that Berry associates with the woods in *The Long-Legged House* and *The Unforeseen Wilderness* often appears in the poetry as a capacity for healing. In "A Standing Ground" (*CP*, 116) he pictures himself escaping anxiety and the fury of argument by climbing up "into the healing shadow of the woods." Berry can find solace in other natural settings, yet he keeps returning to the woods as a symbol of peace and restoration. These can be the familiar woods where he finds refuge "beneath / the blessed and the blessing trees" ("Woods," *CP*, 205) or the unknown "forest of the night" entered by the long hunter, "the true wilderness, where renewal / is found" ("Setting Out," *CP*, 248). In his poetry Berry associates the restorative peace of the woods with stillness and solitude and often with a sense of mystery, but this peace also has to do with the way the woods embody seasonal and longer cycles of renewal. By describing the fall of the year as "time's wound," he suggests the original Fall and a more profound kind of healing than one would associate with nature's annual renewal: "In the household / Of the woods the past / Is always healing in the light" ("The Anniversary," *CP*, 168). As I have suggested, Berry also sees the reclaiming of cleared land by the forest as a form of healing. In the optimistic view of the resilience of the "old forest" in "Window Poems" he can speak of "healed fields / where the woods come back / after time of crops, / human history / done with" (*CP*, 75).

Berry refers frequently in the poetry to a movement beyond ordinary speech and thought into a state of consciousness he associates with harmony with the natural world.[15] This often appears as a movement into silence and darkness that is a form of absence.[16] To be "at home in the world," Berry suggests in the first of two poems entitled "The Silence," one must withdraw "beyond words into the woven shadows / of the grass and the flighty darknesses / of leaves shaking in the wind"; it must be as though one's "bones fade beyond thought" (CP, 111). Then one can hear "the silence / of the tongues of the dead tribesmen" and the "songs" of the natural world. In the second poem Berry wrestles with the writer's dilemma of having to use words to express perceptions that he finds beyond the reach of language: "Though the air is full of singing / my head is loud / with the labor of words" (CP, 156). The poem moves toward an act of renunciation: "Let me say / and not mourn: the world / lives in the death of speech / and sings there." Berry makes poetry here out of the need to disavow language in order to hear the inexpressible song. Withdrawing from speech and entering the darkness becomes an enabling act, a way of escaping the forms of human consciousness associated with language and of attaining a state of heightened perception. In "Woods" (CP, 205) he suggests the paradoxical nature of this state: "Though I am silent / There is singing around me. / Though I am dark / There is vision around me." This singing can be the song of birds and insects, actual sounds of the woods, but more often it appears in the poetry as a generalized song of the earth, music heard in the mind that expresses the dynamism of the created and creating world: "Let the great song come / that sways the branches, that weaves / the nest of the vireo" ("From the Crest," CP, 194).

One must embrace darkness as well as silence to experience the kind of vision Berry associates with harmony with the natural world.[17] In the poetry this darkness comes to symbolize the unknowable and also the necessity of yielding to the ways of the earth and, finally, to a death understood as part of the cyclical rhythm of nature. It can be the darkness of the woods Berry describes in The Unforeseen Wilderness but also a more generalized "dark / of the earth" that suggests the mysteriousness of natural processes. The growth of roots in the dark of the soil offers a kind of meaning and vitality Berry cannot find elsewhere: "There is no earthly promise of life or peace / but where the roots branch and weave / their patient and silent passages in the dark" ("A Standing Ground," *CP*, 116). To "Learn the darkness," as a voice urges in "Song in a Year of Catastrophe" (*CP*, 117), is to "Put your hands / into the earth. Live close / to the ground" but also to accept the death the earth "requires." Serving the dark means serving the earth, not only by tilling the ground but by entering it at death ("Enriching the Earth," *CP*, 110). One ultimately "bow[s] / to the mystery" by accepting one's life as "a patient willing descent into the grass" ("The Wish to Be Generous," *CP*, 114). Such acceptance is more possible if death is seen as joining one to the continuing life of the earth, as Berry imagines it in his "Requiem" (*CP*, 233) for Owen Flood:

> Now may the grace of death
> be upon him, his spirit blessed
> in deep song of the world
> and the stars turning, the seasons
> returning, and long rest.

One of Berry's strengths as an elegist is his ability to find a distinctive kind of consolation in such a merging with the earth, what he calls in "Three Elegiac Poems" (*CP*, 51) a going "into the life of the hill / that holds his peace." There is no hint of Christian resurrection here, rather a sense of resting from labor and becoming a part of the song of the world.

Learning the dark can also mean establishing a sense of communion with the remembered dead that is possible for Berry when he enters the timelessness of the natural world. In "Song in a Year of Catastrophe" (*CP*, 117) this is a matter of looking "behind the veil / of the leaves" and hearing their voices. This sense of communion can come through following a worn path into the woods and feeling that the dead he has loved are a part of the ground he walks ("in blood, in mind, / the dead and living / into each other pass" ["In Rain," *CP*, 268]). At such moments, associated with experience in particular natural settings, the boundary between living and dead appears to dissolve and the continuity with those who have lived and labored before in the place seems strongest. In "The Strait" (*CP*, 246) Berry makes the thrush's notes, heard moving closer and then away in the shadows of the woods, a metaphor for "the world's one song . . . passing / in and out of deaths." Berry presents this song as a summons into the dark, where it paradoxically offers "light" and a sense of the ongoing life of the natural world ("It does not attend / the dead, or what will die"). The thrush's song, elusive as well as hauntingly beautiful, embodies an alluring sense of mystery that Berry often discovers in the natural world and here finds consoling, both a medium of communion with the "never forgotten dead" for whom he sorrows and an encouragement to live.

Not surprisingly, woods figure in key scenes in Berry's fiction, as places associated with revelation and peace and often death. In the final chapter of his early novel *A Place on Earth* (1967), which in the revised version Berry titled "Into the Woods," we see Mat Feltner finally accepting the death of his son in the war as he rests at the edge of a grove of great beeches that have overgrown formerly cleared land.[18] Coming into the radiant "presence" of the place, with brilliant gold leaves falling around him, Mat can lose his grief for the past and feel the continuity and rightness of the natural order.

> He feels the great restfulness of that place, its casual perfect order. It is the restfulness of a place where the merest or the most improbable accident is made a necessity and a part of the design, where death can only give into life. And Mat feels the difference between that restful order and his own constant struggle to maintain and regulate his clearings. Although the meanings of those clearings and his devotion to them remain firm in his mind, he knows without sorrow that they will end, the order he has made and kept in them will be overthrown, the effortless order of wilderness will return.[19]

Fiction enabled Berry to dramatize the tension between the human struggle to maintain a sphere of domestic order and the peace and order he habitually associates with the woods and to show this tension dissolved in a moment of epiphany.

In his novel *The Memory of Old Jack* (1974) Berry again used the woods as the setting for his climactic scene, in this case involving another kind of acceptance. Before Jack Beechum can die he goes into the woods as the day ends to sit under a walnut tree, a place to "rest in and be still." In this state of perfect repose and awareness he can let go or, as Berry imagines it from the perspective of the continuing life of the land, his fields can break free of his "demanding" and his "praise": "He feels them loosen from him and go on."[20] Mat Feltner lets go of his grief, Jack Beechum of the fields he has continued to haunt out of anxiety and habit even after turning over the farming of them to a young couple of whom he thoroughly approves. In both instances Berry suggests that true repose can come only from relaxing the human desire to control one's environment by perpetuating the order embodied in cultivated fields. The natural order to which they yield is "casual," "effortless," inevitable.

In a particularly memorable story, "The Boundary" (published in *The Wild Birds* [1985]), Berry shows Mat Feltner entering the woods again, this time at age eighty, to see whether a fence line that he worries about has been mended. As in *A Place on Earth,* the woods in this story suggest the peacefulness and continuity of the natural world and diminish the importance of the human struggle. The woods of the story, however, take on another dimension as Mat crosses psychic as well as physical boundaries to enter a state in which the dead seem as present to him as the living and the lovely and peaceful woods suggest the appeal of death: "I could stay here a long time."[21] Berry describes Mat, once he has settled into the stillness and attentiveness that the place demands, as "pass[ing] across into the wild inward presence of the place" (82), then allowing himself to drift down along the stream that the fence line follows in order to "prolong his deep peaceable attention to that voice that speaks only of where it is" (84). The dramatic tension of the story is generated by the reader's growing awareness that Mat's journey is taking him—physically, and psychically—beyond his capacity to return. Berry describes Mat at one point as seeming "to be walking in and out of his mind," or in and out of time (90). These woods are dangerous, then, yet Berry makes us see the movement out of time and toward death as a natural and not undesirable process. When Mat nears the limits of his physical strength in his effort to make his way back home, he begins to imagine himself lying down in some inviting place as if to "enter directly into the peaceableness of this place, and turn with it through the seasons, his body grown easy in weight" (95). The scene recalls Berry's imagining of his own death at the end of "A Native Hill," the last essay of *The Long-Legged House,* as he lies "easy" in the October leaves. In the story that follows "The Boundary" ("That Distant Land") Berry makes Mat's slow dying over the course of the summer seem a leavetaking for which he is fully ready and shows those he leaves gradually assimilating into their daily lives the prospect of his death.

Berry's short novel *Remembering* (1988) ends with yet another climactic scene in the woods, in which the protagonist Andy Catlett finally accepts the loss of his hand to a mechanical cotton picker in a farming accident. Like Mat Feltner in *A Place on Earth* and Jack Beechum in *The Memory of Old Jack,* Andy rests in the woods at the base of a tree, in this case deep woods along an old wagon road to Port William that he seeks out after restless travels that have taken him as far as San Francisco. Berry makes Andy's experience of these woods even more revelatory and transforming than the experiences of his earlier protagonists. He risks the device of a dream vision that arises from the "hopeless dark sleep" into which Andy lapses when a "dark man" gives him a mysterious healing touch, as if from the world of the dead. The springtime scene to which Andy awakens in his dream is the sort Berry has represented before—with birdsong, early woodland flowers, and a flowing stream—but it quickly takes on a visionary quality. The trees have become large and old, "as if centuries have passed," and he hears the sunlight singing, "above and beyond the birds' song":

> The light's music resounds and shines in the air and over the countryside, drawing everything into the infinite, sensed but mysterious pattern of its harmony. From every tree and leaf, grass blade, stone, bird, and beast, it

is answered and again answers in return. The creatures
sing back their names. But more than their names. They
sing their being. The world sings. The sky sings back.[22]

The morning light, expressed as music, seems to animate and order
nature, as if embodying a divine presence that evokes an answering
response. But Berry avoids any sense of a transcendent heaven, of
the sort that Milton suggested in *Paradise Lost* in the psalmic morning hymn in which Adam and Eve articulate the praise of the created
world. The songs of the world and of the answering sky form "one
song, the song of the many members of one love" (*Remembering,*
122). Andy, emerging from his personal darkness, sees into the
heart of the natural world and grasps its fundamental joy. When the
strange dark man whose face he never sees leads him out of the
woods to look down on a panorama of Port William and its surrounding fields, he sees a place wholly beautiful—with clean, white
houses among great trees—and wholly joyous, populated by the
remembered and the legendary dead "resting and talking together
in the peace of a sabbath profound and bright" (123). The people
seem to draw their life from the place and from the dynamism of
the enveloping natural world; "[they] are the membership of one
another and of the place and of the song or light in which they live
and move" (123).

This hopeful vision erases the effects of time and loss and
enables Andy to conclude that like the dead of Port William "he lives
in eternity as he lives in time, and nothing is lost" (123). It is Berry's
answer to Milton's vision of a triumphant Christian heaven, as this
is expressed in the lines from *Lycidas* that the dying Mat Feltner
recites to Andy Catlett, without knowing their source, in "That Distant Land":

> There entertain him all the saints above,
> In solemn troops and sweet societies
> That sing, and singing in their glory move,
> And wipe the tears forever from his eyes.[23]

Berry creates a version of heaven on earth in Andy's vision in
Remembering, overriding the familiar Christian opposition of cor-

rupt earth and transcendent heavenly paradise. Berry's distant land ("Oh pilgrim, have you seen that distant land?" Burley Coulter sings in the story of that title) is an idealized version of Port William, arrested in a Sabbath peace. In "Work Song" (published in *Clearing* [1977]) Berry offers a vision of a renewed land, restored and enriched by careful stewardship, in which the river will run clear, an old forest stands, and forgotten springs open: "Families will be singing in the fields, / In their voices they will hear a music / risen out of the ground" (32).²⁴ Berry insists that this is "no paradisal dream" but an ideal attainable through wisdom handed down over generations ("Its hardship is its possibility"). The more rapturous vision of *Remembering* also offers a sustaining ideal—it enables Andy to accept his disability and return to contribute his own kind of help to the community—yet the beautiful land and people he sees united in "one great song" exist in what has become part of a dream that preserves the ideals and the memories of the place, as Berry imagines these. It is a consoling vision, akin to the consolation with which Milton ends *Lycidas,* a perfected version of a deeply familiar and cherished place that holds out the possibility of some kind of permanence.

Woods again figure prominently in Berry's novel *The Life Story of Jayber Crow* (2000), but with a critical difference.²⁵ The fifty acres of woods that Athey Keith has preserved as his "Nest Egg" become the place where Jayber has accidental meetings with Athey's married daughter Mattie, for whom he harbors an austere, undeclared love. This is a woods of big trees, some of them survivors from the time of Daniel Boone, which offers delights of the sort that Berry associates with other special woods, although this one includes periodically flooded bottomland that produces nettles and mosquitoes and must be entered by a path through a cane thicket. Berry shows the woods revealing itself to Jayber and Mattie as they walk through it on those occasions when they happen to meet. They come upon a slope covered with bluebells, a fox that unexpectedly emerges from the undergrowth, "rooms and vistas" that seem arranged for them. The woods offers a kind of calm and satisfaction that one comes to expect from such places in Berry's works. Yet in this case Berry's emphasis has shifted from the presence of the woods itself to its role in bringing Jayber and Mattie together. They walk "in beauty" and silently share their appreciation of the unfolding scene: "The place spoke for us and was a kind of speech. We spoke to each other in the things we saw" (349).

Jayber Crow ends not with a climactic scene of recognition or vision in the woods but with an emotionally charged meeting between Jayber and Mattie in the hospital room where she is dying. The catalyst for the meeting, and for the acknowledgment of unspoken love that it prompts, is Mattie's husband Troy's act of having the woods logged to hold off his creditors a little longer. Berry presents the cutting and hauling of the trees as a brutal business that sickens Jayber and serves as final confirmation of Troy's disregard for the land that his father-in-law, Athey, the prototype of the good farmer, has always nurtured. But his primary emphasis is on the way Jayber's capacity for love and faith survives the ruin of his world. In a novel in which Port William itself is seen to be declining and the noise of the outside world becomes inescapable, the most important truth is the way Jayber is redeemed by his love for Mattie. Berry continues to associate heaven with this world in the novel, but it appears to Jayber as something only glimpsed, "like the reflection of the trees on the water" (351) or the double rainbow he discovers above hilltops glowing in the sun when he emerges from the woods one afternoon after encountering Mattie there.

In the later poetry collected in *A Timbered Choir* Berry articulates more fully than before his sense of an alternative religion associated with meditation in the natural world. The Sunday church bell of the town sends him in a contrary direction, to resume "the standing Sabbath / Of the woods" in the grove of old trees above his farm:

> I go in pilgrimage
> Across an old fenced boundary
> To wildness without age
> Where, in their long dominion
> The trees have been left free. (9)[26]

The woods function as a place where the mind can find rest and the promise of renewal as in the early essays, but the sense of mystery associated with the place has become more explicitly religious:

"Miracle and parable / Exceeding thought" (6). Entering the woods has become a ritual activity, Berry's Sunday morning retreat, and depends upon the recognition that it is "Your Sabbath, Lord, that keeps us by / Your will, not ours" (7). The grove itself has become a natural church, a "high, restful sanctuary" (14) that he describes as a "columned room" where "trees like great saints stand in time, / Eternal in their patience" (65). In the final poem of Berry's first gathering of these poems, *Sabbaths* (1987), the return of the great trees ("Slowly, slowly, they return") suggests the entrance of divinity into the world: "They are the advent they await."[27] The trees become for Berry "apostles of the living light," saying a benediction "over the living and the dead" (*A Timbered Choir,* 95), thus displacing the saints of orthodox Christianity as dispensers of blessings. In *Farming: A Handbook* Berry had imagined the god he "always expected / To appear at the woods edge, beckoning" as a "great relisher of the world" (*CP,* 134). The liturgical imagery with which he ends *Sabbaths* makes a much stronger statement about the immanence of the Christian God in a sanctified natural world.

The worshipful mood that Berry establishes has nothing to do with the anticipation of a Christian heaven. Rather, the sanctuary of the woods "keeps the memory of Paradise" (*A Timbered Choir,* 14). Entering the woods becomes a way of recapturing, at least briefly, a sense of the innocence and vibrance of an unblemished, seemingly timeless natural world "where the world is being made" (*A Timbered Choir,* 35) in the continuous process of creation. One's mind may "move with the leaves" and "live as the light lives" (*A Timbered Choir,* 35) if one enters in a properly detached and receptive state. Berry uses familiar imagery of song, derived here as often in his poetry from the birdsong of the woods, to suggest a state of harmony with the being of nature: "Sabbath economy / In which all thought is song, / All labor is a dance" (*A Timbered Choir,* 56). The difference in the Sabbath poems is that the song of the earth has become psalmic praise of the Creator. The epigraph from *Sabbaths,* which Berry repeats in *A Timbered Choir,* a passage from Isaiah referring to the rejoicing of the fir trees and the cedars of Lebanon, signals the celebratory tone of the collection: "The whole earth is at rest and is / quiet: they break forth into singing" (Isa. 14:7).

In a number of his Sabbath poems Berry takes an optimistic

view of the promise of a "healed harmony" of woods and field, wild and domesticated space, made possible by what he calls "loving work" (*A Timbered Choir,* 14). In one of the earliest, from 1979, field and woods rejoicing together "rejoin the primal Sabbath's hymn" (*A Timbered Choir,* 13), recalling a paradisal ideal of the union of human and natural orders. In the poems from 1991 to 1997 collected in *A Timbered Choir* Berry focuses increasingly on the actual work of farming, particularly in the long poem from 1991 given the title "The Farm." Berry frames his georgic prescriptions in this poem with reminders of the relationship between "timeless woods" and "timely work." He begins by taking the reader through the woods to a point where his farm spreads out below, a vision "seen / As on a Sabbath walk" of the possibility of a life "whose terms / Are Heaven's and this earth's" (136). After many lines of practical advice about such matters as plowing and planting, lambing and haying, he admonishes the reader: "Do not forget the woods." The woods "measure" the field, teaching nature's lessons of preservation and economy, and also establish the larger context in which Berry sees the farm and the work of farming: "The farm's a human order / Opening among the trees / Remembering the woods" (142). Berry urges going to the woods after labor to rest and be reminded of an order independent of human effort that serves as a reminder of what this effort cannot accomplish. One should praise this order and make the land recall "in workdays of the fields, / The Sabbath of the woods" (147). Thus the "Sabbath economy" and the human economy of good work, undertaken in love and gratitude, come together in Berry's vision of the ideal life.

The forest is "mostly dark" in Berry's earlier poetry, something one has to "stay brave enough" to keep entering, whether this is the actual forest with its mysteries and uncertainties or the wilderness of instinct upon which marriage depends.[28] In the poems of *A Timbered Choir* the pull of the woods, represented by the familiar grove Berry seeks out on his walks, has become stronger. These woods seem mostly light, a place where one finds "heavenly work of light and wind and leaf," set off against the "peopled dark / Of our unraveling century" (14). Wildness has become freedom from external control: "wild is anything / Beyond the reach of purpose not its own." The good white oak that Berry admires is "unbe-spoke," not domesticated for human use like an orchard tree but liv-ing in the grove "by its will alone, / Lost to all other wills but Heaven's—wild" (188). Berry concluded *Sabbaths* with an image of fall as "brightened leaves": "we are pleased / To walk on radiance, amazed. / O light come down to earth, be praised!" (*A Timbered Choir*, 83). He ended *The Long-Legged House* with an image of himself lying in newly fallen leaves, accepting the idea of the body's "long shudder into humus" (113). The calm embrace of mortality Berry arrives at in that work gives way in the Sabbath poems to joyful cel-ebration of fall, with the change signaled by "brightened" leaves wel-comed as a manifestation of the splendor of the natural world and of the divinity that pervades and orders it.

If the hopefulness and the religious affirmations of the Sabbath poems collected in *A Timbered Choir* distinguish this book from Berry's previous work, these poems nonetheless reiterate and deepen themes that can be seen emerging in the early essays and poetry and also in the fiction. They show more clearly than ever the importance for Berry of preserving wild places within the human order, sacred groves that can function as places of meditation and of spiritual solace and renewal. And as reminders of the natural order and the limitations of human understanding and capacity. To go into the woods with Berry requires an intense awareness of the natural world and its processes; a respect grounded in humility and awe that can become a form of worship; and a willingness to submit to an order that transcends human works and preoccupations, ulti-mately by accepting the naturalness of death. Berry represents what he calls the "effortless order of wilderness" as always evolving and always "unforeseen": "I go amazed / Into the maze of a design / That mind can follow but not know."[29] He would have us amazed as we go into the woods, continually surprised and humbled, and at the same time drawn by the confidence that only here can one find the kind of design that explains and satisfies.

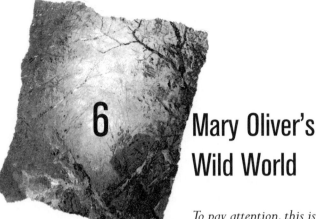

6 Mary Oliver's Wild World

To pay attention, this is our endless and proper work.

——MARY OLIVER, *White Pine*

I am sensual in order to be spiritual.

——MARY OLIVER, *Winter Hours*

Mary Oliver takes extraordinary risks in her poetry to achieve a unique, lyrical vision of what she calls "the wild world" (*BP,* 120). The poetry reveals her passion for this "wild world," a fascination with possibilities for intimacy with it, and a remarkable capacity to "pay attention" to details that reflect the violence and decay as well as the loveliness of wild nature. Oliver characteristically writes about the fields and woods, the ponds and seashore, in the vicinity of her home in Provincetown, Massachusetts. Like Thoreau, she is a walker who returns repeatedly to favorite places, finding wildness in familiar landscapes and looking "deeper, and deeper, into the ordinary," as she has put it.[1] Yet she also responds to the appeal of a wildness associated with nineteenth-century America, particularly in the poetry up to and including *American Primitive,* the 1983 book for which she won the Pulitzer Prize. *American Primitive* offers Oliver's most compelling evocations of the lost worlds of nine-teenth-century America and at the same time celebrates the possi-bility of experiencing a joyous, sensuous pleasure based upon empathy with wild nature, a contemporary kind of American prim-

itivism. It marks an important stage in her evolving search for intimacy and a sustaining joy in her encounters with the natural world.

Oliver presents a gallery of nineteenth-century primitives in
American Primitive: "John Chapman (24–25)," "Tecumseh (77–78),"
"The Lost Children" (12–15), and "Ghosts" (28–30), the first three
set in Ohio and the last on the Northern plains. Her John Chapman
is a saintly figure who submerges personal pain in his work of dis-
persing apple seeds as he wanders through the Ohio forests barefoot, in a sackcloth shirt, sharing hollow logs with the creatures he
finds there. This "good legend," memorialized by "patches / of cold
white fire" in the spring woods, contrasts with the bitter legacy of
the death of Tecumseh. Oliver's retelling of his failed attempt to
preserve Ohio for the Shawnee is an exercise in remembering "the
wounds of the past" and in empathizing with the anger and defiance
of the doomed Shawnee ("I would like to paint my body red and go
out into / the glittering snow / to die"). By dwelling on the fact that
Tecumseh's body was not found and imagining alternate scenarios
to explain this, Oliver suggests that the wounds are still open.

In *The River Styx, Ohio* (1972) Oliver showed "the Indians"
turned into performers and stripped of dignity by the need for
stereotypes. The Indians of "The Indians Visit the Museum" (25)
smile and shake hands, emblems of defeat rather than of the primitive life Oliver prefers to imagine:

> I wish they had come with drums and painted faces.
> I wish they had come ornate and proud
> And like a definition of wild places
> Had merely stood before the crowd.

White Eagle (or "Mr. White") dances in feathers and paint, "[strut-
ting] for money" in Ohio schoolrooms built on the plains where his
fathers lived ("Learning about the Indians"). Oliver returned to this
scene in "Two Kinds of Deliverance" (*Dream Work*), juxtaposing
appealing images of returning spring with a memory of pain that
will not go away: "there flares up out of a vanished wilderness, like
fire, / still blistering: the wrinkled face / of an old Chippewa /
smiling, hating us, / dancing for his life"[2] (61–62). The later poem

offers a more complex response to a troubling memory, associating the renewal of nature and its reminders of a "vanished wilderness" with a history that cannot be elided.[3]

In *American Primitive* Oliver amplifies her view of Native Americans, restoring their dignity and honoring their lives, in Tecumseh's case by conjuring up his angry ghost. In "The Lost Children" (12–15) she presents beguiling images of life in the woods or with Indian captors. These are mainly domestic: Lydia Osborn making a small house of twigs and moss and drinking "the cold water of some / wild stream"; Isaac Zane as an adult rejecting the chance to return to white society and remaining in the house he built by the Mad River with "the beautiful dark woman, / the White Crane, Myeerah." Oliver's concluding reflection on the Wyandot chief's refusal to barter the captured boy, "I like to imagine / he did it / for all of us," insists on the importance of stories that suggest the possibility of a simpler kind of life more attuned to the natural world. An earlier poem, "Aunt Leaf" (*Twelve Moons,* 47–48), shows the speaker temporarily escaping from her own family on journeys into the woods with an invented friend who has the power to change them into "two foxes with black feet." If not a lost child, the speaker in this poem is at least one who wanders and identifies with the lives of wild creatures. Tellingly, she gives her imaginary companion Native American names ("Shining Leaf, or Drifting Cloud / or The Beauty of the Night") and represents her as an "old twist of feathers and birch bark."

The experience of Meriwether Lewis gave Oliver another window on a lost world, in this case the "wild green America" of the Northern plains at the beginning of the nineteenth century. Her poem "Sturgeon Moon—The Death of Meriwether Lewis" (*Twelve Moons,* 35–36) focuses not on the adventures that made Lewis famous but on the effect of his encounter with wilderness, the "fatal exchange / of his life" with the rivers and the "plains full of dark / beasts" that leaves him incapable of functioning in the society to which he returns.[4] Oliver, always conscious of the power of nature to amaze, shows Lewis submitting to the spell of this "green dazzling paradise" and then, in the backwater existence of his life after the expedition, still brooding on "the green man he had left / deep inside Yellowstone." The circumstances of his mysterious early death do not matter to Oliver, only his transformative experience of wilderness.

In *American Primitive* Lewis reappears in "Ghosts" (28–30), Oliver's powerful lament for the buffalo, not as the heroic Lewis overcoming all obstacles including the supreme one of crossing the Bitterroots but as a gentler figure kneeling on the prairie, "near the Bitterroot Mountains," to watch day-old chicks in a "cleverly concealed" sparrow's nest lined with buffalo hair. The scene is an intimate one, suggesting Lewis's attentiveness to the intricacies of the prairie landscape and also the vulnerability of the young chicks, described as having fallen "helpless and blind / into the flowered fields and the perils" of this world. It anticipates the startling conclusion of the poem, in which Oliver recounts a dream of watching a buffalo cow give birth and then kneeling down and asking mother and calf to "make room for me." This is an even more intimate scene, in which the observer seems to be granted a view of the domestic secrets of the natural world. The plains landscape appears both nurturing and immensely inviting, as we see the cow licking and nursing her calf "in a warm corner / of the clear night / in the fragrant grass / in the wild domains / of the prairie spring."

The emotional intensity of this concluding scene, with its powerful suggestion of a need to believe in the continuity of life and the wholeness of a pristine natural world, is heightened by contrast with the elegiac stanza preceding it, in which Oliver imagines a ring of roaring bulls outwaiting the wolves "that are also / *have you noticed? Gone now.*" The stanzas immediately preceding this one evoke the reckless human slaughter of the buffalo and the ghost dances of the Sioux that would bring them back. Oliver's image of rotting carcasses left to stink in the heat embodies anger at the "mutilations" that the Lewis of "Sturgeon Moon" could not imagine as he told his stories of "a world too rich / to be left unmolested." In "Ghosts," as in "Tecumseh," Oliver performs an act of remembrance, countering what she calls in "Tecumseh" (77) the risk of "forgetting what we should never forget," in this case by juxtaposing the violence against the buffalo with scenes that suggest the complex and fragile web of relationships that the presence of the buffalo symbolized.

Oliver's attraction to nineteenth-century America links her with earlier celebrants of wilderness and wildness, including Thoreau, who found the primitive in images of the presettlement life of "the Indians," as well as in woods and swamps—for example, in the "dim and misty" vision of a solitary canoeist gliding up the Millinocket that figures his own sense of a lost world.[5] Oliver's visions of early America register a more acute sense of loss than one finds in Thoreau, or in Audubon, who could lament the anticipated decline of the buffalo even as he and his party contributed to this by their own shooting. They also reflect a different kind of "yearning toward wildness," in Thoreau's phrase, concerned with recovering the capacity for living an intense and richly sensuous life that one finds more fully realized in Oliver's renderings of her own experience, most of it close to home.

Oliver recognizes, as Thoreau did, that one need not travel far to find wildness. A relatively early poem, "Indian Pipes" (*River Styx, Ohio,* 8) marks a "new kind of vision" that erases the child's sense of wildness as something remote from daily life. The experience of finding Indian pipes growing on a nearby hill assumes the force of revelation: "The longer I live the more I sense / Wilderness approaching; I used to walk / Miles to find wild things; now they find me." The poem registers Oliver's growing skills of observation and also a recognition that the "edges" marking the separation of the wild from the familiar were only in her mind ("all's one").[6] Oliver generalizes the lesson of "Indian Pipes" in another poem from the same book, in which "Going to Walden" (53) becomes more than "a green visit": "It is the slow and difficult / Trick of living, and finding it where you are." The evidence of *American Primitive* suggests that Oliver has mastered this trick of living, which involves paying attention to the natural world and learning to recognize and identify with its wildness.

The wildness that Oliver celebrates in *American Primitive* takes many forms, including the imagined wildness of an earlier America. Her observation of the "secret" birth of a fawn and its play in flowery June fields offers a sense of beauty and promise so powerful that it evokes the wish to shed her old life and "to begin again, / to be utterly wild" ("A Meeting" *AP,* 63). The yearning here is for wildness as the dreamlike innocence and harmony with the natural world embodied by the deer. Seeing a fox "explode underfoot" ("Tasting the Wild Grapes," *AP,* 23) and whales leaping in play off the cape ("Humpbacks," *AP,* 60–62) prompts more excited responses, as her emotions are kindled by the unpredictable energies she observes. The moral that Oliver offers in "Humpbacks" ("nothing will ever dazzle you / like the dreams of your body") reflects her preoccupation in *American Primitive* with awakening the body by attuning herself to the instincts and forces she finds in the natural world, as if in this way she could participate in these herself.

The opening poem of the book, "August" (*AP,* 3), is the first of a sequence in which Oliver seems to realize the dreams of her body by identifying with bears, continuing a motif from *Twelve Moons.*[7] In one of several bear poems in that book, "Hunter's Moon—Eating the Bear," she imagines herself eating the bear's flesh and absorbing its power and grace, as well as its breath and hairiness. One is reminded of the attraction of Thoreau and Muir to wild food, particularly of Muir's desire to absorb the wild energies of the Douglas squirrel. In *American Primitive* bears exemplify for Oliver wildness associated with a primitive instinct for sating the appetite. She describes herself in "August" as spending "all day" devouring blackberries, suggesting her kinship with bears by referring to "this thick paw of my life" and "this happy tongue." For Thoreau, eating wild apples and huckleberries was a means of partaking of the vitality of wild nature and of demonstrating his allegiance to the wild as opposed to the civil. The opposite of the wild for him was the domesticated and predictable, in thinking and behavior as well as in the cultivation of fruit. Oliver resembles Thoreau in her dismissal of the "ambition" associated with conventional life, preferring to go off "barefoot and with a silver pail, / to gather blueberries," as she puts it in "Am I Not Among the Early Risers" (*WW,* 7–8). Her primary concern in *American Primitive,* however, is to recover a sense of those impulses that drive the natural world; letting herself become as absorbed as a bear in eating blackberries, and as oblivious to the pressure of time, becomes a way of opening her body up to sensuous pleasure: "all day my body / accepts what it is" ("August," 3).

In "Honey at the Table" (*AP*, 57) Oliver engages the reader in a playful fantasy of retreating from the decorous activity of eating at the table to follow a trickle of honey into the forest as it "grows deeper and wilder, edged / with pine boughs and wet boulders, / paw prints of bobcat and bear." Her voice becomes hypnotic, taking the reader into the domain of wild nature and then into the experience of the bear eating honey at the source:

> deep in the forest you
> shuffle up some tree, you rip the bark,
>
> you float into and swallow the dripping combs,
> bits of the tree, crushed bees—a taste
> composed of everything lost, in which everything
> lost is found.

Oliver's "you float into" suggests a trancelike state in which we can recover the taste of honey in its primitive form, before it is made fit for the table by being refined and packaged, perhaps in a bear-shaped container. Her reference to finding "everything lost" embodies a wish for a more profound kind of recovery, for another life in which sensation is fuller and more intense than anything possible in our self-conscious, civilized existence.

Oliver shifts her focus in "Happiness" (*AP*, 71) to the hidden sweetness of the "honey-house" deep in the tree that the she-bear persistently searches out and to the sheer physicality of eating it: "honey and comb / she lipped and tongued and scooped out / in her black nails." Oliver places herself in this poem as an observer celebrating the bear's ability to give herself up to the pleasure of sating herself on the honey, humming and swaying in the tree as if sleepy or drunk. She figures pure happiness by describing the bear looking as though she would fly down to the meadows like an "enormous bee," escaping her heavy body to "float and sleep in the sheer nets / swaying from flower to flower / day after shining day." Bear and imagined bee sway as if transformed by a state like the "sensual inundation" Oliver advocates in "The Plum Trees" (*AP*, 84).

In the final poem of the group Oliver deliberately crosses over into the world of the bear. Whereas in "Happiness" she preserves an observer's distance ("I saw her let go"), in "The Honey Tree" (*AP*, 81) she imagines herself climbing the tree and eating honey and "the bodies of bees" in a frenzy of joy. The poem celebrates the moment at which she embraces her own sensuality, as she embraces the physicality of the world ("how I love myself at last! / how I love the world!"). Her body can now sing "in the heaven of appetite" because she has so completely yielded to its urgings. To "love the world," as she proclaims she does, is to plunge into the sensuous pleasure it offers.

Oliver expresses a nostalgia for the most primitive state of all in "The Sea" (*AP*, 69–70), that of a fish in the "motherlap" of the sea from which human life emerged. The bones of the swimmer long to dive "and simply / become again a flaming body of blind feeling." This is the purest sensuality Oliver imagines in *American Primitive*, using her characteristic imagery of fire ("flaming") to suggest an intensity that seems to concentrate the energy of the natural world. This energy becomes explicitly erotic in some poems, most notably "Blossom," "Music," and the concluding poem of the book, "The Gardens." In the April night of "Blossom" (*AP*, 49–50), "there's fire / everywhere: frogs shouting / their desire, / their satisfaction." The poem presents these impulses as natural and irresistible, a "thrust / from the root / of the body" like the thrust from the root of the skunk cabbage, "stubborn and powerful as instinct" ("Skunk Cabbage," *AP*, 44). Here and elsewhere Oliver recognizes the self-consciousness that separates humans from the rhythms of the natural world—we "know" that "we are more / than our hunger"—yet she insists that "we belong / to the moon" and are driven by our "burning" to hurry down "into the body of another."[8]

In "Music" (*AP*, 72–74) Oliver imagines herself metamorphosing into the goat god Pan and producing an irresistibly seductive music. She again uses fire ("it was late summer when everything / is full of fire and rounding to fruition") and the pull of the moon to suggest the body's susceptibility to the forces of the natural world, but her focus is on the power of Pan's wild music to "[light] up the otherwise / blunt wilderness of the body." The full moon and the flaring lightning contribute to creating a setting charged with sexual excitement in "The Gardens" (*AP*, 85–87), in which Oliver shows herself crossing the visible garden to go "down / into the garden / of fire." In the erotic scene with which the poem concludes the land-scape of the lover's body merges with that of the natural world, in a journey seen as "leading deeper into the trees, / over the white fields, / the rivers of bone." Hurrying "into the body of another" here becomes a way of seeking the pleasures and the mystery ("the unknowable / center") of a wildness in which human and natural seem to merge.

In *American Primitive* Oliver associates sexual desire and all the desires of the body with the natural world, through sensitivity to the forces she perceives as driving this world, identification with the instinctive behavior of animals, and acts of engagement such as eat-ing wild foods, tempting happiness into the mind "by taking it / into the body first, like small / wild plums" ("The Plum Trees"). Yet for all of Oliver's efforts in *American Primitive* to show how wild nature can light up the "wilderness of the body"—more central to the book than her recreations of the wilderness of Meriwether Lewis or that of the Eastern forests into which the lost children disappear—she is after something more than a simple celebration of sensuality. In the title essay of her recent book *Winter Hours* (1999), she says, "I am sensual in order to be spiritual."[9] Oliver's concern with what she calls "the spiritual side of the world" and with her own spiritu-ality is more explicit in the later work, but one can see this emerg-ing in *American Primitive*.[10] Her own description of the book sug-gests that the search for pleasure has other dimensions: "*American Primitive* I wanted to be a listing of many perceptual joys. But joy that doesn't end in pleasure. Rather, pleasure that leads to a sense of humility, and a sense of praise, and a sense of mystery, and a sense of wonder."[11]

In the short poem "May" (*AP*, 53) Oliver describes responses stimulated by spring blossoms that "storm out of the darkness:" "the bees / dive into them and I too, to gather / their spiritual honey."

The dynamic of light and darkness that gives this poem its force is fundamental to Oliver's work and suggests how high the stakes can be in her engagement with nature. Oliver's belief in the resilience and the essential goodness of the natural world and in her own capacity for a happiness grounded in the body defines itself in opposition to a darkness that can suggest, among other things, the doubts of the night ("Morning at Great Pond," *AP*, 46–47); the violence of nature (the owl of "In the Pinewoods, Crow and Owl," *AP*, 9, is the "bone-crushing prince of dark days"); and fear of the change that signals our mortality ("the black river of loss" of "In Blackwater Woods," *AP*, 82–83). In "May" belief takes the form of certainty that "the flourishing of the physical body . . . is as good / as a poem or a prayer, can also make / luminous any dark place on earth." Wild creatures, like the egrets who open their wings and step "over every dark thing" ("Egrets," *AP*, 19–20), offer the simplest illustration of the "faith in the world" Oliver seeks. Sustaining such belief for herself can involve struggle to understand and accept the troubling aspects of the natural world, as in "Crossing the Swamp" (*AP*, 58–59), where the primal "black, slack / earthsoup" threatens to engulf mind as well as body but finally suggests the possibility of regeneration. It can also involve the kind of qualified affirmation with which Oliver concludes "Fall Song" (*AP*, 18), that "everything lives . . . forever / in these momentary pastures." As she says in "In Blackwater Woods," we must "love what is mortal" and yet "when the time comes . . . let it go."

Oliver urges her readers to "love the world," but she recognizes that nature can be deadly. Winter drives life to the edge, as in "Cold Poem" (*AP*, 31). A violent storm brings howling wind and lightning that evokes mingled fear and excitement as it tears through "the dark / field of the other" ("Lightning," *AP*, 7–8). A bobcat leaps from the pines with "lightning eyes." Hungry bluefish pour "like fire" over the minnows that are their prey. Such potentially destructive forces are as natural and as worthy of praise for Oliver as the whale's exuberance or the bear's appetite for honey. The capacity for excitement that distinguishes Oliver's response to wildness, whether this is joyful or erotic or charged with fear, can be seen as her way of resisting dullness and what she calls the "paralysis" of death.[12] In *American Primitive* Oliver's connection with the "other" of nature often seems kinetic, a response to a "blazing" that invites a corresponding emotional intensity. Such a response depends upon believing that nature is vibrantly alive, a "country / of original fire" where everything "throbs with song" ("Humpbacks"), and that she can experience its energies and immerse herself in its physical pleasures. Oliver explores the possibilities for vision and for the lyrical expression of rapture more fully in subsequent poetry, but *American Primitive* is remarkable for the way she enlarges our sense of the meaning and value of wildness by registering its impact on her emotional life and, at her most optimistic, declaring her willingness to "believe in everything" ("Morning at Great Pond").

Oliver's sensitivity to what she calls "the infallible energies" of wild nature and her capacity for excitement and rapture make her seem closer to Muir than to Thoreau in some respects.[13] Both Oliver and Muir use imagery of lightning and electricity to express their sense of the natural world as intensely alive, and both respond strongly to the felt energy of storms (Oliver describes trees tossing in the wind as "full of electricity now").[14] Oliver registers a broader range of emotions than Muir, of course, and shows a fascination with the violent energies of predators entirely missing in his work. One finds in Oliver as in Muir, however, a reciprocity between animated observer and the animate nature each perceives. Their own sense of being intensely alive is charged by the kinds of vitality they observe, Muir in the movements of the whole landscape ("everything is flowing") and Oliver in the force and spirit she finds in plants as well as animals.[15] Each blurs the line between animate and inanimate, distinctions Oliver says she doesn't care about. She insists that "the blue bowl of the pond, and the blue bowl on the table, that holds six apples, are all animate, and have spirits."[16] Where Muir finds rocks living, even electric, Oliver asks why we should not think of them as having souls: "What about all the little stones, sitting alone in the moonlight?" ("Some Questions You Might Ask," *HL,* 1)[17]

Divinity does not permeate Oliver's wild world in the way it does Muir's, and her sense of the process of relating to this world is considerably more complicated than his. Muir reconciled the seemingly contradictory aspects of nature—harsh rock faces and destructive storms, on the one hand, alluring mountain meadows on the other—by embracing the paradox that nature is both violent and tender. His confidence in divine love manifesting itself through an ongoing process of creation meant that he could see even the most violent forces as working to enhance the beauty and appeal of nature. Storms might beat down his "sky gardens," but they would emerge fresher and more vigorous. Oliver's oppositions are not so easily resolved and typically involve a more complex dynamic. Light is set off by darkness and may seem to triumph over it, but such victories usually seem precarious and temporary. Nature for her offers thorns as well as nectar. Love, as she frequently reminds her readers, involves risk and pain.

In her many poems about predators (including foxes and snapping turtles, herons and egrets, sharks, seabirds, and especially owls), Oliver makes her readers confront the violent death and the terror that she sees as fundamental to the natural world. She forces us to recognize the "cruel mystery" of this world in "Turtle" (*HL*, 22–23), by showing the snapper taking the teal's "soft children," and is capable of such nightmarish images of death as "flapping, blood-gulping crows" on the ice ("Crows," *HL*, 75). Yet Oliver evokes such horrors to make us recognize that they are part of the natural cycle of life in a world "in which the owl is endlessly hungry and endlessly on the hunt" (*BP*, 20). In "Nature" (*HL*, 55–56) she takes comfort in "the things of darkness / following their sleepy course": "even the screams . . . were as red songs that rose and fell / in their accustomed place." Where Muir reads the "savage" face of a mountain landscape as manifesting divine love, Oliver embraces the paradox that owls can be deadly as well as dazzling in their beauty and efficiency. In fact, she admires the hunting skills of predators as examples of nature's perfection.[18]

The snowy owl of "White Owl Flies Into and Out of the Field" (79–80), the concluding poem of *House of Light* (1990), is beautiful ("like an angel, / or a buddha with wings") and graceful as it strikes and then rises from the field. The extraordinary image

with which this poem ends transmutes death into something amazing and welcome, as Oliver asks us to imagine ourselves carried:

> to the river
> that is without the least dapple or shadow—
> that is nothing but light—scalding, aortal light—
> in which we are washed and washed
> out of our bones.

Here the imagined darkness of death dissolves in an image of
purification and release through light that is "scalding," as if
embodying all the hot energy of the natural world. Oliver's transformation of the river of death into a purifying river of light represents her boldest development of the dominant strand of imagery in
The House of Light.

Light, whether associated with the sun or with what Oliver
calls the "fire of the world," functions as an energizing force and a
sign of the kind of aliveness she craves. It may be associated with the
sudden perception of the snowy egret "tossing a moment of light" in
its wings ("What Is It?" *HL*, 26–28) or the pleasure of "lounging in
the light" listening to the mockingbird ("The Gift," *HL*, 36) or the
recognition of the power of a lion rising from the grass "like a fire,"
with sunlight in his hair, as if emerging into being ("Serengeti," *HL*,
61–63). Oliver's assertion at the end of "The Ponds" (*HL*, 59) associates light with a kind of beauty and positive energy that transcend
the evidence of decay she sees in the lilies:

> I want to believe I am looking
>
> into the white fire of a great mystery.
> I want to believe that the imperfections are nothing
> that the light is everything—that it is more than the
> sum
> of each flawed blossom rising and falling. And I do.

Oliver's embrace of light reflects a capacity for wonder and a readi-
ness to believe in the resilience and the fundamental goodness of the

natural world. Such affirmations are critical to the strain of optimism that runs through her poetry, yet they contend with a sense of
threat, often imaged by darkness, and with a consciousness of the
boundaries that divide her from the natural world.

In her poetry since *American Primitive* Oliver's sense of the possibilities for vision and rapture has continued to evolve, and she has also
shown a more complicated awareness of the ways that self-consciousness and an attachment to language separate us from the life
of nature. A sense of separation is apparent in earlier poems such as
"Entering the Kingdom" (*Twelve Moons,* 21), in which Oliver imagines herself rebuffed by the crows whose "kingdom" she has
intruded upon ("They know me for what I am. / No dreamer, / No
eater of leaves"). In "This Morning Again It Was in the Dusty Pines"
(*NSP,* 23–24) she shows an owl turning away "in disgust" as she tries
to imagine a conversation but fails ("what is it I might say . . . that
would have any pluck and worth in it?"). Words have no place here,
Oliver seems to be saying, only admiration for the indifferent owl as
it turns its "hungry, hooked head" toward her and then away and
"glides / through the wind / like a knife."

In "Whispers" (*Dream Work,* 29–30) Oliver suggests the
difficulty of achieving the kind of ecstasy she renders in "The Honey
Tree" and other poems in *American Primitive.* The later poem focuses
on the "resistance" that prevents her from finding the pleasure she
imagines "shining like honey, locked in some / secret tree" and
"[sliding] into / the heaven of sensation." The poem is driven by the
force of her yearning to be a part of what she observes but feels isolated from, the "reasonless," easy lives of creatures finding happiness
in "the sleek, amazing / humdrum of nature's design . . . to which
you cannot belong." The resistance that Oliver describes in "Whispers" may reflect a particular emotional state, but the sense of separateness from wild nature that she describes here has become a
recurrent theme in her work. What engages Oliver in "The Turtle,"
also from *Dream Work* (57–58), is the "old blind wish" to climb the
hill and lay eggs in the sand each spring. The turtle cannot see herself "apart from the rest of the world"; she is "a part of the pond she
lives in." Oliver makes this unthinking fusion with the natural envi-

ronment seem enviable and at the same time unattainable by the human observer.

In *House of Light* Oliver explores what separates her from the natural world in "Lilies" (12–13) and "The Lilies Break Open Over the Dark Water" (40–41). The former moves from the impulse to live like the lilies in the fields to a sense of the difficulty of forgetting oneself and quieting the mind to the recognition of loneliness in a world where lilies melt "without protest" on the tongues of cattle and the hummingbird "just rises and floats away" from trouble. Yearning develops into a sense of inescapable difference. The recognition is similar but more complex in "The Lilies Break Open," which Oliver begins by evoking the rank fertility of the "flecked and swirling / broth of life" in which the waterlilies "quicken" each summer and then "tear the surface" and "break open" like a natural force. These lilies are not pretty or humanized in any way, rather "slippery and wild." Again Oliver focuses on the difference between herself and the lilies, she standing on the shore "fitful and thoughtful, trying / to attach them to an idea—some news of [her] own life," they "devoid of meaning," simply doing what their "being" impels them to. Her concluding line, "And so, dear sorrow, are you," suggests the pain inherent in the struggle to overcome the gap between our language-bound consciousness and the primal world of the lilies, a pain that seems even more acute in the recognition of the limitations of her role as "imaginer" at the end of a more recent poem, "The Osprey" (*WW*, 21–22).[19]

Oliver asks in "The Notebook" (*HL*, 44–45), "How much can the right word do?," playing on the difference between her preoccupied "scribbling and crossing out" and the turtle's experience of the pond scene she is trying to capture, the turtle who "doesn't have a word for any of it." Reflections on her role as poet and on the inadequacy of language have become more frequent and more subtle over the course of Oliver's work. They figure prominently in *West Wind* (1997). In an early poem from that volume, "Stars" (13–14), she writes of the need to quiet the noise of language in the head, "to be empty of words," in order to "listen" and appreciate the silences of the natural world. The poem suggests the difficulty of connecting in the way she would like despite such efforts. Joy and peace "almost" find her, and at the end of the poem we see her looking up at the stars and "the pure, deep darkness" and offering "one hot sentence after another," a measure of desire rather than of satisfaction.

In "Forty Years" (*WW*, 26) Oliver contrasts her long absorption in the process of inscribing words on paper with the elusive richness of the world she has been trying to render:

> And again this morning as always
> I am stopped as the world comes back
> Wet and beautiful I am thinking
>   That language
>
> Is not even a river
> Is not a tree is not a green field. . . .

Oliver has a talent for making poetry out of her sense of the poet's limitations, offering a detached, sometimes ironic view of her own obsessive activity. In the title piece of *West Wind* (49), she describes herself at her midnight desk "rewriting nature / so anyone can understand it," trying to account for the passion she associates with roses tossing in the wind. The prose poem "Toad" (*WP*, 38) offers a more bemused view of a one-sided conversation with a toad that sat like a Buddha, unblinking, "as the refined anguish of language passed over him."

Oliver's self-consciousness about nature's resistance to her attempts to rewrite it, like her sensitivity to its sometimes alarming energies and its frequent reminders of mortality, makes her continuing efforts to write a visionary poetry seem far from naive. These efforts reflect a persistent quest to render the joy to be found from what Oliver describes in *Blue Pastures* (1995) as "standing *within*" the world's "*otherness*," its "beauty and mystery."[20] In *American Primitive* this joy takes the form of sensuous ecstasy made possible by attuning oneself to the wildness and the pulsing energies of the natural world and letting the body "[accept] what it is." Oliver's imagery of floating and of "singing in the heaven of appetite" in that book, in poems in which she identifies pleasure with the bear's search for wild honey, suggests a rapture dependent upon suspending thought and yielding to the pull of the senses.

The urgency of appetite gives way to a more generalized love of the physical world in "Spring" (*House of Light,* 6–7), a later poem in which Oliver identifies with the happiness exemplified by a she-bear awaking in early spring to ramble down the mountain "breathing and tasting." Her assertion of identity with the bear of this poem is a qualified one that allows for the recognition that her life is also poems and music and cities. Oliver's concluding association of the bear's "perfect love" with "wordlessness," like her earlier reference to "the silence / of trees," signals an awareness of difference. Our self-consciousness and dependence upon language make such unreflective, "perfect" love of the world unlikely, but it can stand as an ideal for Oliver. In "Roses, Late Summer," also from *House of Light* (66–67), her sense of difference becomes explicit. Here the yearning that one sees in "Spring" takes the form of envy of an "unstinting happiness" associated with the natural world:

> I would be a fox, or a tree
> full of waving branches.
> I wouldn't mind being a rose
> in a field full of roses.
>
> Fear has not yet occurred to them, nor ambition.
> Reason they have not yet thought of.
> Neither do they ask how long they must be roses, and
>   then what.
> Or any other foolish question.

The final line mocks the human tendency to ask "foolish" questions, exemplified by the series of questions with which the poem begins, and at the same time implicitly acknowledges that we can't help asking such questions.

Oliver's poetry is more convincing because of her recognition of the difficulty of shedding self-consciousness and the habits of thought embedded in language. It is most remarkable, however, for its affirmations of the possibilities for joy and rapture. Those who learn to "pay attention" to the natural world and to love it, she implies, can enjoy moments when this world reveals itself. Her poems about encounters with deer suggest the possibility of finding joy in unexpected moments of intimacy, the most striking of which is the one she describes in "The Pinewoods" (*WP*, 13–14) in which two deer walk toward her and one touches her hands.[21] She has been "exalted" ever since, she claims, "stalled in the happiness / of the miracle." The seemingly miraculous experience of intimacy she describes contrasts with an attitude she attributes to Thoreau and Audubon in one of the "Excerpts" she offers in *Blue Pastures*: "The danger of people becoming infatuated with knowledge. Thoreau gassing the moth to get a perfect specimen. Audubon pushing the needle into the bittern's heart" (51). Oliver distances herself here from the naturalist intent on possessing the object of study. The revelations she seeks depend upon the patient attentiveness and tact she displays in her encounter with the deer. What she would preserve, through poetry, is the moment of intimacy: "I am still standing under the dark trees, / they are still walking toward me."

In her more recent poems Oliver seems increasingly interested in capturing visionary moments. By placing her "favorite story" of Baucis and Philemon at the center of "Mockingbirds" (*WP*, 16), she suggests the possibility of a miraculous revelation, like that of the two visitors who in leaving reveal themselves to be gods. They bless the old couple for their humility and receptiveness, "their willingness / to be attentive." By ending the poem with an image of herself listening to the mockingbirds, opening the "dark doors" of her soul ("I was leaning out; / I was listening"), Oliver suggests that she enjoys her own revelatory moment. What matters here is the attitude, one of alertness and receptivity to something as extraordinary as a visit from emissaries of another world. The prose poem "December" (*WP*, 51) offers a momentary glimpse of what Oliver calls the "real world" in the form of a deer with leaves growing from its antlers who steps from the woods and then vanishes: "The great door opens a crack, a hint of the truth is given—so bright it is almost a death, a joy we can't bear—and then it is gone." We need to believe in the possibility of such revelatory moments, Oliver suggests in "I Looked Up," in *White Pine* (1991, 54): "What wretchedness, to believe only in what can be proven."[22] The miracle in this poem is the ability to see the bird as "lighting up the dark branches of the pine" and rising "wreathed in fire," with "wings enormous and opulent." The vision it describes expresses Oliver's sense of joy and wonder at the reassuring intensity of the natural world ("What misery to be afraid of death").

Oliver's ability to see the bird "wreathed in fire" is related to her ability to smell the fragrance of a white pine, in the prose poem of that title that concludes the book, and believe that "Everything is in it." By moving from this fragrance to that of the opened body of an African antelope Oliver reprises one of the more striking images in Thoreau's "Walking," giving it a startling twist. For Thoreau the antelope's "perfume" of trees and grass, like the odor of muskrat in the trapper's coat, symbolizes the wildness he would have his readers assimilate by becoming "so much a part and parcel of nature" that their smell advertises their natural haunts. Oliver makes the encounter with the fragrance of the antelope a liminal moment that is dangerous as well as pleasurable: "The fragrance is almost too sweet, too delicate, too beautiful to be borne. It is a moment which hunters must pass through carefully, with concentrated and even religious attention, if they are to reach the other side, and go on with their individual lives" (*WP*, 55). This moment, as Oliver renders it, offers a transcendent experience so overpowering that it threatens one's grasp on ordinary life. The book ends with the image of Oliver having finished her walk but "just standing, quietly, in the darkness, under the tree," as if unwilling to leave.

Oliver addresses the nature of vision in one of the central poems of *West Wind,* "The Rapture" (32–33), in which she describes herself as perceiving "the shimmering, and the extravagant" as she wanders the summer fields. The capacity to see the "extravagant," which gives Oliver the sensation of rising above the earth ("once or twice, at least"), depends upon the attentiveness and willingness to believe that she talks about in earlier poems. She focuses here, however, on the "passion" that calls her forth and strips her "clean," making happiness possible. For Oliver as for Berry, vision requires moving into a condition of harmony with the natural world after shedding the distractions of everyday life and quieting the "noise" of language. Where Berry seeks a calm and beauty in the woods that he associates with the natural order of creation, however, Oliver pursues "rapture the gleaming, / rapture the illogical the weightless——." This is an excited pursuit that finds rapture in an earth seen as "heavy and electric." The concluding lines of "The Rapture," "At the edge of sweet sanity open / such wild, blind wings," insist upon the irrational, "wild" character of the transcendent experience.

Oliver begins *West Wind* by invoking two of her masters in the art of visionary poetry: "all eternity / is in the moment this is what / Blake said Whitman said" ("White Butterflies," 3). In the concluding poem of the book, "Have You Ever Tried to Enter the Long Black Branches" (61–63), Oliver claims that "once in a while, I have chanced, among the quick things, / upon the immutable."[23] A confidence that she can glimpse an immutable "real" world, her version of finding eternity in the moment, runs through the spiritual adventuring of Oliver's recent poetry. The door to the inner chamber opens, a bird flashes in the trees, fields shimmer. By associating truth with such visionary moments, Oliver may seem to put great pressure upon herself and her readers to experience these moments and the rapture they promise. Yet she mainly urges attitudes of the sort she finds in Whitman's *Leaves of Grass* ("attention, sympathy, empathy") and a kind of seeking she finds necessary to a "luminous" life.[24] Such seeking implies that ordinary life, whether embodied by domesticity or by ambitious striving, is not enough. For Oliver, being alive involves opening the window of the soul and venturing into the world with an alertness to the possibility of vision.

Oliver sounds rather like Whitman at times in "Have You Ever Tried to Enter the Long Black Branches," especially in her urgent questions and exhortations:

> Well, there is time left—
> fields everywhere invite you into them.
>
> And who will care, who will chide you, if you wander away
>   from wherever you are, to look for your soul?
>
> Quickly, then, get up, put on your coat, leave your desk!

Oliver has written about her strong attraction to Whitman in several recent essays, praising *Leaves of Grass* for qualities that one can recognize in her own poetry, including passion, affirmation, risk, and an attraction to the erotic and the mystical.[25] She has clearly learned from Whitman's use of what she calls "the familiar, the intimate, the companionable voice" and his technique of challenging the reader with provocative questions, as well as from other poetic models she acknowledges, among them the "intensified vernacular" of James Wright.[26] She echoes Whitman's diction ("I swear") and develops her own versions of some of his preoccupations (with invoking the soul, embracing leisure, celebrating sensuous pleasure). The influence of Whitman seems stronger than ever in Oliver's latest book, the extraordinary long poem *The Leaf and the Cloud* (2000), with its frequent catalogues and repetitions, its frank declarations, and its extended celebrations. There are reminders of Whitman's democratic instincts ("I will sing for the iron doors of the prison, / and for the broken doors of the poor"), but Oliver's empathy is primarily for the things of the natural world: the hummingbird, the moth, the "pale lily." If Oliver has learned from Whitman's stylistic habits and his bold voice, her own confident voice cannot not be mistaken for anyone else's.[27]

One suggestive point of connection with Whitman can be seen in Oliver's increasing tendency to invoke fields and grass as emblems of the appealing wildness of the natural world and objects of contemplation. Oliver's attraction to fields as the locus of vision in "The Rapture" and "Have You Ever Tried to Enter the Long Black Branches" is anticipated by earlier poems, including "The Summer Day" (*HL,* 60), in which she presents herself as strolling the fields all day and falling and kneeling in the grass, as a demonstration that she knows "how to be idle and blessed." One is reminded of Whitman again here, particularly of lines from the opening of "Song of Myself" that Oliver has quoted: "I loafe and invite my soul, / I lean and loafe at my ease . . . observing a spear of summer grass."[28] Oliver finds value and meaning in the grass, as well as refuge (in the prose poem "Grass," *WP,* 17), and a sense of mortality ("Do you adore the green grass, with its terror beneath?" [*NSP,* 22]). In "Have

You Ever Tried to Enter the Long Black Branches" lying or sitting in the grass becomes a way of entering the natural world, or at least of trying to enter it: "Never to lie down in the grass, as though you were the grass! / To sit down, like a weed among weeds, and rustle in the wind!" The effort to become like the grass, or like weeds rustling in the wind, implies a willingness to let go ("Leave your desk!") and seek whatever truth might be found in harmony with the natural world. Oliver makes this a matter of urgency that takes precedence over the demands of success and social conformity, displaying a rebelliousness that aligns her with Thoreau as well as with Whitman.

In the first new poem in *New and Selected Poems* (1992), "Rain" (3–7), Oliver asks "What shall I do, what shall I do?" and then answers her question with a declaration:

> I do not want anymore to be useful, to be docile, to
>   lead
> Children out of the fields into the text
> Of civility, to teach them that they are (they are not)
>   better
> than the grass.

Oliver's subversiveness recalls Thoreau's ("Give me a wildness whose glance no civilization can endure"), but her embrace of wildness has more to do with the capacity for vision and joy than with an assault on conventional values. Civility for her represents a decorum that must be shed if she is to become alive to the world she perceives. In the short poem "Rice" (*NSP,* 38), which recalls the earlier "Honey at the Table," she urges:

> I don't want you just to sit down at the table.
> I don't want you just to eat, and be content.
> I want you to walk out into the fields
> Where the water is shining, and the rice has risen.
> I want you to stand there, far from the white table-
>   cloth.
> I want you to fill your hands with the mud, like a bless-
>   ing.

The fertile mud, like the "dripping comb" of "Honey at the Table," suggests the real, vital world that Oliver opposes to the domesticated one represented by the white tablecloth. As in the earlier poem, she takes the reader on a quest for origins, restoring a lost connection with the natural world. This is how to understand and experience "rice" (or "honey"), she seems to be saying.

In another poem from the same period, "A Certain Sharpness in the Morning Air" (*NSP*, 41–42), Oliver recalls "thinking of the old, wild life of the fields when as I remember it, / I was shaggy, and beautiful, / like the rose." In her imagination ("as I remember it") Oliver blends into this "old, wild life of the fields," assuming its innocence, despite the odor of skunk ("the brushing of thorns") that dominates the poem and serves as a reminder of pain and darkness. If the "wild life" that Oliver imagines here represents an unobtainable ideal, the effort to recover a sense of it can nonetheless offer a form of salvation. In "Poppies" (*NSP*, 39–40) the world's "roughage" temporarily "shines like a miracle / as it floats above everything," suggesting the possibility of transcendence. The "earthly delight" that she finds in the "bright fields" of poppies offers "a kind of holiness, / palpable and redemptive," and serves as a stay against darkness (here the "deep, blue night"). This renewal of happiness among the poppies becomes an answer to the consciousness of mortality that haunts this poem as it does many others.

The kind of spiritual seeking that produces visionary moments and the rapture that can accompany them has become increasingly important in Oliver's poetry. Such seeking dominates *The Leaf and the Cloud,* in which Oliver offers affirmations of poetry and of the beauty of the physical world but keeps returning to questions about meaning. In "Work" (*LC,* 14) Oliver moves beyond a preoccupation with nature's resistance to her efforts to capture it to affirm her commitment to words and to suggest a way of reconciling the poem and the world:

> So I will write my poem, but I will leave room for the world.
> I will write my poem tenderly and simply, but
>   I will leave room for the wind combing the grass,
> for the feather falling out of the grouse's fan-tail,
> and fluttering down, like a song.

She can assert that "glory is my work" and say, of the poem as a whole: "I have placed one word next to another / to build something full of praise" (*LC*, 49). Yet she includes a section about death ("Gravel"), in which she claims that it is the nature of humans, of all ages and social stations, to ask the frenzied question: "what does it mean, that the world is beautiful— / what does it mean?" (*LC*, 42).

Although *The Leaf and the Cloud* fundamentally affirms life and Oliver's role as poet ("This is the world, and this is the work of the world"), it also registers her uncertainties and the tension she feels between "the skeptic and the amazed—" (*LC*, 25). She introduces the possibility of belief in God but without finding answers to her insistent questions, identifying with "the doubters running over the hot fields, / crying out for faith" (*LC*, 50). She expresses her willingness to believe and adore but stops short of declaring her belief and offers her praise obliquely, through a catalogue of perceptions that suggest the freshness and wonder of the world: "The first streak of light in the darkness, / the first bird to sing, / the first whale to rise out of the black water" (*LC*, 50). Oliver balks at the notion of a Christian heaven, imagining herself becoming a part of the natural world she knows:

> Listen, I don't think we're going to rise
> In gauze and halos.
> Maybe as grass, and slowly,
> Maybe as the long-leaved, beautiful grass. (*LC*, 44)

She can affirm the vitality and beauty of the world, symbolized for her by a field of grass growing "wild and thick" and shedding pollen "like a mist full of life," but she preserves a sense of ambiguity about what we are to believe: "We stand at the edge of the field, sneezing. / We praise God, or Nature, according to our determinations" (*LC*, 10).

The seeking that characterizes Oliver's more recent poetry proceeds from a belief in the soul that encompasses all life ("I believe in the soul—in mine, and yours, and the bluejay's, and the pilot whale's"), and it involves exploring and writing about "landscapes in which we are reinforced in our sense of the world as a mystery."²⁹ Recognizing and honoring a nature understood to be beyond our comprehension, whether represented by pinewoods or fields or the "wild waste spaces of the sea," takes the place of religious ritual for Oliver: "As with prayer, which is a dipping of oneself toward the light, there is a consequence of attentiveness to the grass itself, and the sky itself, and to the floating bird. I too leave the fret and enclosure of my own life. I too dip myself toward the immeasurable."³⁰ Oliver introduces a form of prayer in *The Leaf and the Cloud* with a series of blessings ("Bless the tongue / for it is a maker of words") and what could pass for a liturgy in her repetitions of the phrase "This is the world," as if substituting the body of the world for the body of Christ.³¹ Her movement toward more formal religious expression in the poem is most striking in the alleluiahs of swans and moth with which she concludes: "alleluiah alleluiah / sighs the pale green moth / on the screen door" (*LC,* 52), an expression of praise that implies acceptance of the conditions of life in this world. Earlier in the poem we see the "delicate" wings of the moth juxtaposed with the beak of the crow.

If Oliver seeks rapture in a natural world that she finds mysterious, she looks for it in the familiar and the obvious, the "manifest." Particular visionary moments are less important than the attitude with which one approaches the natural world: "What I mean by spirituality is not theology, but attitude" (*Winter Hours,* 102). One can see such "attitude" in characteristic stances or activities: watching at the edge of a pond; sitting or lying in the grass; standing "amazed" or expectant ("For there I am, in the mossy shadows, under the trees"); above all walking.³² Oliver, like Thoreau, numbers herself among the "long-distance walkers" and sees walking as a means to discovery. She describes it in *Winter Hours* (1999) as the way she chooses to live, one that she sees as common to the "mystically inclined": "I am forever just going out for a walk and tripping over the root, or the petal, of some trivia, then seeing it as if in a second sight, as emblematic" (99).³³ In "Am I Not Among the Early Risers" Oliver shows herself beginning each day in a spirit of affirmation, "as I [step] down from the porch and set out along / the green paths of the world." Despite an awareness of pain and uncertainty that the poem reflects, she insists that she is still able to find "in every motion of the green earth . . . a hint of paradise," at sixty as she did at twenty.

Oliver's walking embodies an optimism about her capacity to find joy in the natural world that underlies her injunctions to her readers. By showing herself walking through the fields and woods, over the dunes and along the beach, she invites the reader to join her in the emotional risk-taking that her form of adventuring implies. To abandon "caution and prudence," as she urges in "Have You Ever Tried to Enter the Long Black Branches," means for Oliver to open oneself up to the unpredictable wildness of the natural world and the possibility of chancing upon "the immutable." *West Wind,* like Thoreau's "Walking," ends with a turning homeward at sunset:

> Now the sun begins to swing down. Under the peach-
>   light,
> I cross the fields and the dunes, I follow the ocean's
>   edge.
>
> I climb. I backtrack.
> I float.
> I ramble my way home.

Where Thoreau ends on a rising note by recapitulating his image of walking as a crusade ("So we saunter toward the Holy Land"), invoking the possibility of a transforming vision ("a great awakening light"), Oliver portrays herself as simply rambling home through a familiar landscape. Yet her quiet assertion, "I float," reminds the reader that she finds her own kind of transcendence in walking. The word suggests the freedom from anxiety and the burden of self-consciousness that Oliver associates with creatures (the humming-bird that "floats away" in "Lilies," the "seven dancers floating" of "Seven White Butterflies") and also the rapture made possible by an alert, intimate engagement with the natural world.

Conclusion

Near the end of "A Native Hill," the last of the autobiographical essays that make up *The Long-Legged House,* Wendell Berry describes a downhill walk along a familiar stream (Camp Branch) that becomes an excursion into history as he meditates on his own past and that of the place. Old stonework that created walled springs and pools reminds him of older, slower rhythms of labor and of the patience that made such labor possible. The erosion of "virgin soil" loosened by plows suggests a more material kind of loss, of the country as it appeared before white settlement. The stream leads Berry down to an old barn, the last of his "historical landmarks," where he crosses a significant boundary: "I stoop between the strands of a barbed-wire fence, and in that movement I go out of time into timelessness. I come into a wild place."[1] As we have seen, Berry is aware of the costs of insisting upon a dichotomy between nature and culture. Experience has taught him that the wild and domestic can coexist (as in the hedgerows and other "margins" that he describes in his essay "Preserving Wildness"). Yet he repeatedly returns in his writing to the experience of crossing a boundary into a heightened awareness of the "other" world of nature. One can find similar boundary crossings in most of the other writers I discuss, even if the boundaries are not always so clearly delineated as they tend to be for Berry. In our concern with showing how nature is marked by culture and with challenging the habit of opposing the two, we run the risk of losing sight of the central importance for writers of the sort that I discuss here of understanding and representing wildness as different, other. They find wildness in various kinds of places and apprehend it in ways conditioned by individual and cultural differences, but they are all seekers of one kind or another, and this seeking can take the form of a spiritual quest for truths that they do not find in the social and cultural orders that they inhabit.

John James Audubon stands out from the other writers that I consider by his complicated self-presentation and by the nature of his quest. His seeking was manifestly physical, a form of hunting that was driven by ambition for fame and financial security, as well as by a passion for the birds he sought to record. His desire to possess these birds by shooting and then painting them often seems in conflict with his pleasure in observing their behavior. Yet his enthusiasm for the natural scenes he describes comes through strongly, despite the moderating effects of MacGillivray's editing of his prose. Audubon's capacity for wonder was fed by a sense that in his many forays he was escaping from the tedium of business and family responsibility into a natural world of great richness and drama. Beyond the frontier regions of settlements and squatters' cabins lay what he saw as the primal American wilderness. This was for Audubon the theater in which he acted his role of American Woodsman, presenting himself as the heroic naturalist venturing into the "dark depths" of forest and swamp and surviving whatever physical and psychological trials they might offer. Wilderness could appear to him sublime; Edenic in its luxuriant growth; or, in the case of Labrador, "terrifyingly wild." Audubon embraced his adventures in the spirit of an explorer bringing back reports of unknown lands and little known species. At the same time he lamented the impending loss of much of what he celebrated. The nostalgia that he frequently expresses reveals a sense of fatalism about the pace of change that masks his own complicity in this change. Although he protested some of the practices he observed, particularly in his later travels, he accepted and occasionally celebrated American expansionism and saw himself as recording a wilderness that was inexorably receding.

Henry David Thoreau attempted, with incomplete success, to leave the domesticated world behind in his excursions to the Maine woods. He relished the wildness that he found there, symbolized for him by Tom Fowler's beer, which he imagines as distilling the "most fantastic and spiciest sprays of the primitive wood, and whatever invigorating and stringent gum or essence it afforded."[2] Yet he perceives wilderness in relation to the civilization he sees as impinging on and ultimately displacing it. He preserves something of the detachment of a travel writer reporting his observations and comes to identify with the "poet" he imagines as not really belonging in these wet woods but drawing strength and inspiration from them on periodic visits. In his habitual walks in the vicinity of Concord, shaking off village thoughts and seeking "the springs of life," Thoreau enacts a more characteristic kind of psychological boundary crossing. While he was highly conscious of and interested in human influences on the landscape, he habitually sought out evidences of "primitive" nature. His "yearning" for wildness manifested itself throughout his writing career in such urges as a craving for wild food and a desire to participate in the life of a marsh by wading into it. Thoreau was a passionate observer of natural processes who sought an intimacy with wild nature beyond that of an observer, yet he recognized the elusiveness of such intimacy. His representation of himself in "Walking" as someone living a "border life" who makes "transient forays" into "another land" than the one described by owners' deeds expresses a sense of the limits of his capacity to know the "Nature" to which he was so strongly drawn. The fact that he thought of himself as crossing a border and, shifting his metaphor, as seeing the bounds of known farms fade and the picture underneath emerge suggests that for him the other world of nature and the truths it could reveal were ultimately knowable only through some kind of transforming vision.

Thoreau's practice of walking can be understood as an expression of faith in the potential of engagement with wild nature to revitalize the spirit and liberate the mind. By presenting walking as analogous to Christian pilgrimage (in "Walking"), he makes it seem a quest for truth. By the end of "Walking" this has come to seem a pilgrimage toward a Holy Land that reveals itself only in visionary moments such as the one he enjoys in the "pure and bright" light of a November sunset that can transform a meadow to a "paradise." The moment is unexpected and fleeting, yet Thoreau imagines it recurring, as the natural phenomenon repeats itself "forever and ever" and cheers others in the future. When he turns toward home, he sees the west-facing slopes and woods gleaming "like the boundary of Elysium." This boundary, like that of his imagined Holy Land, is one that he can cross only in the imagination, but for him the scene holds the promise of a "great awakening light" that will light up whole lives.

Without Thoreau and Emerson, it would have been harder for John Muir to articulate his own religion of nature. Although he was much more disposed than Thoreau to use religious language uncritically as a way of praising the divinity he saw acting through nature and much less inclined to recognize barriers to participating in the natural world, he was enabled by Thoreau's sense of the value and power of wildness. The most remarkable aspect of Muir's wilderness writing is his emphasis on the dynamism of nature, and this springs from a conviction that he could not only perceive its energies but experience them. His epiphanies tend to come at times when he feels himself most in tune with nature's "flowing" and released from any sense of constraint. When he describes himself as drifting "enchanted" and seemingly weightless through high Sierra meadows, with the feeling that his body is "all one tingling palate" (*MFS,* 153), his sense of liberation seems as much physical as intellectual. Muir's prose is most rapturous at such moments of identification with nature.

While the introduction to the mountains that Muir describes in *My First Summer in the Sierra* proceeds by stages, he gives the sense of crossing a boundary when he reaches the edge of a great pine forest at twenty-five hundred feet and feels that he has at last truly entered the mountains and shed the claims of the dusty world below ("We are now in the mountains and they are in us" [*MFS,* 16]). The conversion experience that he goes on to describe enables him to yield to the "pulsing" rhythms that he senses in the natural world, suspending his sense of the need to "save time" and giving himself up to the heightened sensuous pleasures he experiences.

Where Thoreau's claim that the swamp is his *sanctum sanctorum* is an act of provocation, intended to unsettle orderly habits of thought, Muir's celebration of the glories he found in his mountain temples is a call to worship. As a prophet of the gospel of wilderness, a self-proclaimed John the Baptist, he offers a vision of "holy mountains" and psalmic waterfalls that depends upon excluding any sense of impurity. To sustain his vision Muir needed to minimize the effects of human activity on the landscape and to emphasize the contrast between a world in which the splendors of divine creation were still unfolding in a dynamic process of change and a drab lower world dominated by economic concerns. His insistence upon this dichotomy helps to explain why Muir has become a problematic figure for those who are increasingly concerned with the porousness of the boundaries between nature and culture.

Edward Abbey's rendering of wilderness, in his case the desert Southwest, is far more conflicted and ambiguous than Muir's. His best work, *Desert Solitaire,* continues to be read partly because he consistently questions his own efforts to capture in language what he describes as "the elemental and fundamental, the bedrock" (*DS,* 6). Yet despite his doubts about such efforts and his tendency to undermine the visionary moments that he allowed himself, Abbey persisted in regarding the desert as a distinctly "other" world where space and time assumed different dimensions. His satire is directed primarily at those who lack sufficient understanding of and respect for what he saw as unique about the desert landscape. Entering this landscape and its special places often involves a sense of boundary crossing for Abbey. In *Desert Solitaire* he describes the feeling of dislocation that he experiences in his initial drive to Arches and his ranger station via a twisting dirt road, on a windy night in which eerie shapes appear in the headlights; he and his friend Waterman can enter the "terra incognita" of the Maze only by making their way down a barely passable jeep trail and then rappelling over a cliff edge. Such journeys, like the one he and his friend Newcomb take through Glen Canyon on their trip down the Colorado River, convey Abbey's sense of entering a strange and different world, often challenging and in some way mysterious.

Abbey's questing is in its simplest sense a form of adventuring, in a familiar American tradition. In exploring a corner of the Maze, rafting down wild rivers, and taking solo hikes in the most forbidding desert terrain he could find he was demonstrating his toughness, what he would call his "desert rathood." The guiding spirit of his trip through Glen Canyon is John Wesley Powell, and he invokes the mountain men (Jim Bridger, Jedediah Smith) who preceded him in this country, as well as the Mormon pioneers who spent four months getting their wagons across the canyon with what he represents as heroic perseverance. Yet Abbey clearly was seeking more than adventure. What engages him most in Glen Canyon is the dreamlike character of the river trip and the "fantastic" appearance of the canyon itself. Such places provoke Abbey to speculate on the mysterious and the supernatural and then, in a reflexive reaction against the tendency to such speculation, to reaffirm his commitment to the dogmatic reality of rock and sun that says nothing and means nothing but itself. This tension reflects Abbey's sense that our perception is inevitably subjective, limited by our language and its associations, and also his responsiveness to the appeal of a landscape that seems to promise glimpses of the unknowable. Abbey recognized his need for the immutable ideal that the desert represented for him, and he wanted it to retain the mystery that its silence and aura of timelessness evoked. He responded to the challenge of exploring the Maze, but one senses that he did not want to reach its "unknown heart" even if he could have.

Both Wendell Berry and Mary Oliver are capable of writing in a visionary mode and yet aware of the limits of vision. They are conscious of how what Berry calls the "labor of words" can separate one from the natural world with which they seek intimacy in their different ways. Both, however, describe visions or visionary moments in which they achieve a heightened state of awareness. Berry addresses questions that he recognizes as religious, often adapting religious language, while rejecting the opposition of heaven and earth that he finds in organized religion. He offers an alternate religion, especially in the Sabbath poems collected in *A Timbered Choir,* that involves coming into the "presence" and the "mystery" of the natural world, with its "nonhuman time." For him the highest state of awareness involves movement through silence to a consciousness of the singing of nature, an expression of vitality and joy analogous to the "pulsing" of the natural world that Muir describes.

Berry regularly looks to the natural world as a source of solace and healing and a refuge from anxiety and the contentions of public life. His epiphanies typically involve a recognition of the "casual, perfect order" and the enduring character of the natural world and a corresponding recognition of the limits of human understanding that argues a need for humility. Berry's sense of the "presence" of the natural world reflects a confidence that it is ordered and animated by divinity, which suggests affinities with Muir and even with the more conventional piety of Audubon, for whom the cheering notes of a wood thrush after a rainstorm suggest divine agency. He imagines a past of "big woods" and a future in which the big trees will eventually reclaim cleared land, by predictable stages. But Berry's vision comprehends the human community as well, characterized by the fellowship of its members and a commitment to "loving work" that establishes a "harmony between forest and field."[3] His ideal is a concord of workday and Sabbath, field and forest, that recognizes the possibility of understanding daily life and work as compatible with a sense of belonging to a larger, natural order.

Oliver believes in the "otherness" of wild nature, which constitutes for her a "real world" with energies and a contagious joyousness that she does not find elsewhere, although her boundary crossings are not so clearly demarcated as Berry's. She is more inclined to focus on the difficulty of participating in a natural world from which her self-consciousness and her awareness of "rewriting nature" separate her, although she insists upon the possibility of visionary moments in which she temporarily overcomes her sense of separation. Oliver pays more attention to the nature of vision than any of the other writers that I discuss, suggesting that this depends upon a capacity to believe in more than "what can be proven," to "cast aside the weight of facts" and "float a little / above this difficult world" ("The Ponds," *HL,* 59). One must be disposed to look for the transcendent, recognizing that it may appear only in glimpses. She can see a bird "wreathed in fire" in a pine tree because she believes in the possibility. As she says at one point: "If you think daylight is just daylight / then it is just daylight" ("Sand Dabs, Three," *WW*, 24).

It makes sense to speak of Oliver's spirituality, as it does of Berry's, although she is more likely to subvert the language of Christianity than to adapt it, at least until the poetry of *The Leaf and the Cloud*. She substitutes paying attention for prayer and rejects notions of sin and penance. The happiness that she finds in "bright fields" of poppies becomes her own "kind of holiness" ("Poppies," *NSP*, 40). Such happiness always begins with the senses for Oliver, but sensuous pleasure increasingly seems insufficient for her. In *The Leaf and the Cloud*, playing on a passage from Ruskin's *Modern Painters* describing the veil between humans and the "burning light" of heaven that she uses as an epigraph, she expresses her commitment to a poetry of vision:

> I will sing for the rent in the veil.
> I will sing for what is in front of the veil, the
> Floating light.
> I will sing for what is behind the veil—
> Light, light, and more light.[4]

However Oliver's search for meaning has evolved and continues to evolve, a fascination with light has been a constant in her poetry. Her world gleams, shimmers, and blazes with light, suggesting the intensity she finds in natural phenomena as well as their promise of a truth that she finds both desirable and elusive. One could say that she is always seeking "more light," paying close attention to the manifest in order to discover the transcendent.

Not all writers who display a "yearning toward wildness," to invoke Thoreau once more, are so concerned with discovering the transcendent. They may be attracted to the color and swirl and sexual energy they find in natural phenomena (like Pattiann Rogers in her poetry) or to the promise of healing and refuge they find in familiar natural settings (like Terry Tempest Williams in *Refuge*). They may be more drawn by a sense of freedom or aliveness or by the pleasures of learning to appreciate rhythms and intimations of order that take them outside familiar social and domestic routines. They may respond to evidences of wildness in an urban environment; to landscapes in which they can read the record of previous cultural usage; or to remote, relatively undisturbed natural preserves. What is remarkable is the persistence of the desire to experience wild nature, for whatever purposes and in whatever places. This desire seems likelier to increase than to diminish and to continue to give rise to new ways of imagining wildness.

Notes

Introduction

1. William Bradford, *Of Plymouth Plantation* (New York: Random House, 1952), 62.

2. Ann Ronald, ed., *Words for the Wild* (San Francisco: Sierra Club Books, 1987). In a poll by *Sierra* magazine of favorite environmental books (March/April 2001) Muir trailed only Aldo Leopold, with *My First Summer in the Sierra* his most popular book.

3. See, for example, Ramachandra Guha's "Radical American Environmentalism and Wilderness Preservation" (1989) and "Deep Ecology Revisited" (1998), in *The Great New Wilderness Debate,* ed. J. Baird Callicott and Michael P. Nelson (Athens: University of Georgia Press, 1998), 231–45, 271–79. For a critique of Guha's perspective, see Philip Cafaro and Monish Verma, "For Indian Wilderness," in *The World and the Wild,* ed. David Rothenberg and Marta Ulvaeus (Tucson: University of Arizona Press, 2001), 57–63. See David Rothenberg, introduction to *The World and the Wild,* xi–xxiii, for a spirited defense of the idea of wilderness that is sensitive to recent criticisms.

4. William Cronon, "The Trouble with Wilderness; or, Getting Back to the Wrong Nature," in *Uncommon Ground: Toward Reinventing Nature,* ed. William Cronon (New York: W. W. Norton, 1995), 69–90; Michael Pollan, *Second Nature* (New York: Atlantic Monthly Press, 1991), 189–90. For the most comprehensive discussion of the evolution of American attitudes toward wilderness see Roderick Nash, *Wilderness and the American Mind,* 3d ed. (New Haven: Yale University Press, 1982). Max Oelschlaeger, in *The Idea of Wilderness* (New Haven: Yale University Press, 1991), offers a philo-

sophical perspective on the wilderness tradition in America, with attention to significant literary texts.

5. Susan G. Davis, *Spectacular Nature* (Berkeley: University of California Press, 1997); Jennifer Price, *Flight Maps* (New York: Basic Books, 1999).

6. Price, *Flight Maps,* xix.

7. See William Cronon, *Changes in the Land: Indians, Colonists, and the Ecology of New England* (New York: Hill and Wang, 1983), and Shepard Krech III, *The Ecological Indian: Myth and History* (New York: W. W. Norton, 1999). William M. Denevan describes extensive alteration of the landscapes by Indian activity in North and South America and makes the important point that many of the eyewitness accounts of a "pristine" America date from a time (1750–1850) when Indian populations had been drastically reduced by disease and alterations of the landscape were less visible than they were in 1492. William M. Denevan, "The Pristine Myth: The Landscape of the Americas in 1492," *Annals of the Association of American Geographers* 82 (1992): 369–85.

8. See Thomas Vale, "The Myth of the Humanized Landscape," *Natural Areas Journal* 18 (1998): 231–36. Vale argues that it makes more sense to think of the precontact landscape as a mosaic of pristine and humanized conditions (humanized to varying degrees) and demonstrates that the humanized portions of the present Yosemite National Park constitute a relatively small percentage of its total area.

9. Michael Soulé, "The Social Siege of Nature," in *Reinventing Nature: Responses to Postmodern Deconstruction,* ed. Michael E. Soulé and Gary Lease (Washington, D.C.: Island Press, 1995), 137–70.

10. Gary Snyder, "Is Nature Real," in *The Gary Snyder Reader* (Washington, D.C.: Counterpoint, 1999), 388. An earlier version of Snyder's essay appeared in the winter 1996–97 issue of *Wild Earth,* which was largely devoted to refutations of Cronon's "The Trouble with Wilderness." The aggressiveness of some of the pieces in this issue reflects concern about the political costs of devaluing the term "wilderness."

11. Lawrence Buell, *The Environmental Imagination: Thoreau, Nature Writing, and the Formation of American Culture* (Cambridge: Harvard University Press, 1995). Buell builds upon pioneering work in the field, including John Elder, *Imagining the Earth* (Urbana: University of Illinois Press, 1985); Peter Frizzell, *Nature Writing and America: Essays upon a Cultural Type* (Ames: Iowa State University Press, 1990); Scott Slovic, *Seeking Awareness in American Nature Writing* (Salt Lake City: University of Utah Press, 1992); and Sherman Paul, *For Love of the World: Essays on Nature Writers* (Iowa City: University of Iowa Press, 1992). Buell enlarges his focus to include such issues as environmental justice and the representation of urban nature in his recent book, *Writing for an Endangered World* (Cambridge: Harvard University Press, 2001). Jonathan Bate has spurred the development of ecocriticism in Britain

with *Romantic Ecology: Wordsworth and the Environmental Tradition* (New York: Routledge, 1991) and his recent *The Song of the Earth* (Cambridge: Harvard University Press, 2000).

12. Cheryll Glotfelty and Harold Fromm, eds., *The Ecocriticism Reader* (Athens: University of Georgia Press, 1996). Also see especially the summer 1999 special issue of *New Literary History* on ecocriticism, with an afterword by Lawrence Buell, "The Ecocritical Insurgency," 699–712.

13. See Kent Ryden, "Big Trees, Back Yards, and the Borders of Nature," and John Tallmadge, "Resistance to Urban Nature," in *Reimagining Place,* ed. Robert E. Grese and John R. Knott, *Michigan Quarterly Review* 40 (winter 2001; special issue), 126–40, 178–89.

14. For some of these, see Leonard M. Scigaj, *Sustainable Poetry* (Louisville: University Press of Kentucky, 1999), 131.

15. Peter Friederici, *The Suburban Wild* (Athens: University of Georgia Press, 1999), meditates on his experience of remnants of wildness near his childhood home in a northern suburb of Chicago and argues that wildness "can exist comfortably in our cities and suburbs" (6).

16. Jack Turner, *The Abstract Wild* (Tucson: University of Arizona Press, 1996), 83–84.

17. Conservation biologist Donald M. Waller argues this point and defines wildness in terms of an organism's or habitat's "ecological or evolutionary context, that is, its habitual relationships to other organisms and the surrounding environment" (546–47). Like other conservation biologists he would shift the argument for wilderness preservation to such grounds as biodiversity and ecosystem health. Donald M. Waller, "Getting Back to the Wrong Nature," in *The Great New Wilderness Debate,* ed. J. Baird Callicott and Michael P. Nelson (Athens: University of Georgia Press, 1998), 540–67.

18. Cronon, invoking Snyder on the omnipresence of wildness, concludes "The Trouble with Wilderness" with an exhortation to learn to honor the wild instead of "imagining that we can flee into a mythical wilderness to escape history" (89–90). But this opposition depends on a relatively narrow construction of wildness and ignores the fact that many who are attracted to wildness, including Snyder, would see it as embodied most fully by what they understand as wilderness. Snyder describes wilderness as "a *place* where the wild potential is fully expressed, diversity of living and nonliving beings flourishing according to their own sorts of order." Gary Snyder, *The Practice of the Wild* (San Francisco: North Point Press, 1990), 12.

19. For the text of the act see J. Baird Callicott and Michael P. Nelson, eds., *The Great New Wilderness Debate* (Athens: University of Georgia Press, 1998), 120–30. The definition section of the act goes on to specify various conditions.

20. Henry David Thoreau, *Excursions,* vol. 5 of *The Writings of Henry David*

Thoreau, ed. Bradford Torrey and Francis H. Allen (Boston: Houghton Mifflin, 1906), 226.

21. For a particularly wide-ranging and provocative exploration of the meanings of wildness, drawing upon Eastern as well as Native American thought, see Snyder, *Practice of the Wild.* Turner, *Abstract Wild,* consciously echoes Thoreau in arguing for a more absolute commitment to wildness than he finds in managers of wilderness areas and most proponents of wilderness. See especially chapter six, "In Wildness is the Preservation of the World."

22. Henry David Thoreau, *The Maine Woods,* ed. Joseph J. Moldenhauer (Princeton: Princeton University Press, 1972), 54. I am indebted to Steven Fink's discussion of this passage in *Prophet in the Marketplace* (Princeton: Princeton University Press, 1992), 173–74.

23. Snyder, *Practice of the Wild,* 14.

24. Aldo Leopold, *A Sand County Almanac* (1949; Oxford: Oxford University Press, 1966), 98.

25. Leopold, *Sand County Almanac,* 101.

26. Wendell Berry, *Collected Poems, 1957–1982* (San Francisco: North Point Press, 1984), 137–38.

27. Edward Abbey, *Desert Solitaire* (New York: Ballantine, 1968), 218. Abbey praises Thoreau for his capacity to convey a sense of wonder: "Living a life full of wonder—wonderful—Henry tries to impart that wonder to his readers." Edward Abbey, *Down the River* (New York: E. P. Dutton, 1982), 28.

28. Mary Oliver, "When Death Comes," in *New and Selected Poems* (Boston: Beacon Press, 1992), 10.

29. Mary Oliver, "Death at a Great Distance," in *House of Light* (Boston: Beacon Press, 1990), 43.

30. Thoreau, *Excursions,* 217.

31. Thoreau, *Excursions,* 226.

32. *CP,* 69.

33. Mary Oliver, "Am I Not Among the Early Risers," in *West Wind* (Boston: Houghton Mifflin, 1997), 8.

34. Thoreau, *Excursions,* 205.

35. Mary Oliver, "A Meeting," in *American Primitive* (Boston: Little Brown, 1983), 63.

36. Barry Lopez, *Arctic Dreams* (New York: Charles Scribner's Sons, 1986), 257.

Chapter 1

1. John James Audubon, *Ornithological Biography,* 5 vols. (Edinburgh: Adam and Charles Black, 1831–39), 5:vi.

2. Adam Gopnick comments on the "uncanny intensity" of Audubon's art, "its haute-couture theatricality and ecstatic animation, its pure-white backgrounds and shadowless, cartoonish clarity," and sees him as influenced by the "artificial, theatrical style" of David and committed to the "hard-edged draftsmanship of French neoclassicism." Adam Gopnick, "Audubon's Passion," *The New Yorker,* 25 February 1991, 96, 98.

3. Michael Branch sees Audubon, like Bartram and Wilson before him, as "the romantic type of the solitary wanderer" and observes that he cultivated the sense of the naturalist as romantic hero. Michael Branch, "Indexing American Possibilities: The Natural History Writing of Bartram, Wilson, and Audubon," in *The Ecocriticism Reader,* ed. Cheryll Glotfelty and Harold Fromm (Athens: University of Georgia Press, 1996), 282–302.

4. See Byron's account of Boone and the "sylvan tribe of children of the chase, / Whose young, unawaken'd world was ever new." Lord Byron, *Don Juan,* in *The Poetical Works of Lord Byron* (London: Oxford University Press, 1945), canto viii, stanza lxv.

5. Alice Ford, *John James Audubon,* rev. ed. (Norman: University of Oklahoma Press, 1964; New York: Abbeville Press, 1988), 175–76.

6. As Scott Russell Sanders has shown in the excellent introduction to his *Audubon Reader* (Bloomington: Indiana University Press, 1986), 1–17, Macgillivray preserved the "shape and texture" of Audubon's prose while fixing problems of syntax and polishing the diction to bring it more in line with current standards of propriety. Sanders sees the prose as becoming more British and less colorful and idiosyncratic in the process (introduction, 2–3). Audubon's granddaughter Maria edited the journals she published (1826–29, 1833, 1843) much more severely, to support her contention that "there is not one sentence, one expression, that is other than that of a refined and cultured gentleman." Maria R. Audubon, introduction to *Audubon and his Journals,* 2 vols. (1897; New York: Charles Scribner's Sons, 1994), 1:ix–x. Fortunately, letters and journals for several years are available in unaltered form. See Howard Corning, ed., *Letters of John James Audubon, 1826–1840,* 2 vols. (Boston: The Club of Odd Volumes, 1929); Howard Corning, ed., *Journal of John James Audubon Made during his Trip to New Orleans in 1820–21* (Boston: The Club of Odd Volumes, 1929); Alice Ford, ed., *The 1826 Journal of John James Audubon* (Norman: University of Oklahoma Press, 1967); Howard Corning, ed., *Journal of John James Audubon Made While Obtaining Subscriptions to his Birds of America* (Cambridge, Mass.: Business Historical Society, 1929).

7. Ford, *John James Audubon,* 21ff. For the most thorough examination of the evidence regarding Audubon's birth and his early years in France, see Francis Hobart Herrick, *Audubon the Naturalist,* 2 vols. (New York: D. Appleton and Co., 1938; New York: Dover Publication, 1968).

8. In a reminiscence uncovered and published by his granddaughter

Maria, Audubon describes his father as making excursions to Louisiana, where he married a beautiful and wealthy "lady of Spanish extraction" who moved with him to "Sainte Domingue" after Audubon's birth. Maria R. Audubon, *Audubon and his Journals*, 1:7.

9. Alice Ford, author of the definitive biography, finds no evidence in Jacques Louis David's records that he did.

10. Herrick, *Audubon the Naturalist*, 1:61.

11. See Frank Levering, "The Enchanted Forest," in *The Bicentennial of John James Audubon*, ed. Alton A. Lindsey (Bloomington: Indiana University Press, 1985), 76–95, for a discussion of Audubon as storyteller. Levering relates Audubon's exaggerations to the vogue of the tall tale. Herrick details the inaccuracies of a number of the episodes. See Herrick, *Audubon the Naturalist*, volume 1, chapter 18.

12. Despite his disparagement of Wilson's findings, Audubon was capable of pirating details—even drawings (the female redwing, the female Mississippi kite)—from *American Ornithology*. See Peter Matthiessen, *Wildlife in America* (New York: Viking, 1987), 118–19.

13. Christoph Irmscher has commented on Audubon's efforts to erase the boundary between author and reader to give his observations of bird behavior a sense of immediacy. See *The Poetics of Natural History: From John Bartram to William James* (New Brunswick, N.J.: Rutgers University Press, 1999), 200–202.

14. I base my observations upon examination of selected manuscripts in the collections of the Beinecke Library of Yale University and the libraries of the American Museum of Natural History and the New York Historical Society and of photocopies of manuscripts held by the Smithsonian Institute. Scott Russell Sanders based his conclusions about Macgillivray's revisions upon a comparison of the published version of "Pitting of the Wolves" with the manuscript version reproduced by Alice Ford in *Audubon, By Himself* (Garden City, N.Y.: Natural History Press, 1969); he concludes that the revisions preserve "the essential vision of the text" (Sanders, introduction, 2–3). On the basis of a comparison of Audubon's manuscript account of the ruby-throated hummingbird with the published version, Christoph Irmscher concludes that Macgillivray's revisions made the prose more readable without losing "the tone, the structure, and the central images" of the text (*Poetics of Natural History*, 199–200). I would add that Macgillivray frequently omits words and sentences, presumably to save space and make the prose clearer and more efficient, although his omissions and substitutions also have the effect of moderating Audubon's enthusiasm. While Macgillivray eliminates some of Audubon's literary mannerisms (periphrases, for example), he introduces many of his own, such as "finny" to modify Audubon's "food." He prods Audubon at times, as in a penciled note on the brief account of the

water ouzel: "I do not like the above article, it should be *remodelled* intirely, being too *uncertain* as to species etc."

15. *OB,* 4:43. The manuscript is in the Beinecke Library.

16. The manuscript ("The Anhinga or Snake Bird") can be found in the Special Collections Department of the Smithsonian Institutions Libraries.

17. Irmscher comments on Audubon's effort to justify this action and on the "strange mixture of release and remorse" he finds in Audubon's accounts of shooting birds. See *Poetics of Natural History,* 208–9.

18. The manuscript of Audubon's account can be found in the American Museum of Natural History in New York. Audubon occasionally offers an apologetic aside—for example, after he describes shooting at a female anhinga on her nest: "It was cruel thus to disturb her in her own peaceful solitude; but naturalists, alas! seldom consider this long, when the object of their pursuit is in their view and almost within their grasp" (*OB,* 4:151). Chris Beyers explores Audubon's expressions of guilt over shooting birds and his efforts to rationalize his actions. See "The Ornithological Autobiography of John James Audubon," in *Reading the Earth,* ed. Michael P. Branch, Rochelle Johnson, Daniel Patterson, and Scott Slovic (Moscow: University of Idaho Press, 1998), 119–28.

19. Amy R.W. Meyers sees this image as reflecting Audubon's own role as predator. See her essay, "Observations of an American Woodsman: John James Audubon as Field Naturalist," in *John James Audubon: The Watercolors for "The Birds of America,"* ed. Annette Blaugrund and Theodore E. Stebbins (New York: Random House, 1993), 48. Irmscher also comments on this image and, at more length, on Audubon's struggles with the captive golden eagle that he drew. See *Poetics of Natural History,* 225–30.

20. See Meyers, "Observations of an American Woodsman," 49–53. Meyers offers readings of the tale of the doves' courtship and that of the bluejays' untroubled enjoyment of eggs stolen from the nests of other birds.

21. Michael Harwood and Mary Harwood attribute Audubon's anthropomorphizing to his sense of a connection with the birds he observed. See "In Search of the Real Mr. Audubon," *Audubon Magazine* 87 (1985): 115.

22. William Bartram, *Travels* (1791; New York: Viking, 1988), 65.

23. Chateaubriand, who did not get beyond the East in his travels in America, also imagines a startling array of animal life, including bears drunk on wild grapes and caribou bathing in lakes. François-René de Chateaubriand, *Atala and René,* trans. Rayner Heppenstall (1801; London: Oxford University Press, 1963), 4. Theodore E. Stebbins Jr. notes the possible influence on Audubon of Chateaubriand's *Atala,* which went through twelve editions between 1801 and 1805, and his companion romance *René.* See "Audubon's Drawings of American Birds, 1805–38," in *John James Audubon: The Watercolors for "The Birds of America,"* ed. Annette Blaugrund and Theodore E. Stebbins (New York: Random House, 1993), 3–26.

Chateaubriand describes the hero of *René* entering the "marvellous wilderness" of Kentucky on a hunting trip.

24. Bartram evokes a mood of enchantment even more freely, sometimes several times on a page. He describes himself as "seduced by these sublime enchanting scenes of primitive nature" (*Travels,* 107). Chateaubriand describes the teeming animal life of Louisiana as spreading "l'enchantement et la vie." François-René de Chateaubriand, *Atala,* ed. Raymond Bernex (Paris: Editions Bordas, 1968).

25. Stephen Greenblatt, *Marvellous Possessions* (Chicago: University of Chicago Press, 1991), 25.

26. The backgrounds to Audubon's bird paintings were frequently supplied by assistants, initially by Joseph Mason and later mainly by George Lehman and Maria Martin.

27. See Barbara Novak, *Nature and Culture,* rev. ed. (New York: Oxford University Press, 1995), chapter 3, for a discussion of the way the eighteenth-century sense of the sublime was absorbed into "a religious, moral, frequently nationalist concept of nature" (38).

28. Maria R. Audubon, *Audubon and His Journals,* 1:386, 1:390, 1:403.

29. Audubon portrays the daily life of the expedition in the episode "Labrador" (*OB,* 3:584–87), at the end of which he describes his party as increasingly anxious to leave "the dreary wilderness of grim rocks and desolate moss-edged valleys." Audubon also gives an account of the trip in a letter to his son Victor. See Corning, *Letters of John James Audubon,* 1:240–47.

30. Leo Marx, "Pastoral Ideals and City Troubles," in *Western Man and Environmental Ethics,* ed. Ian G. Barbour (Reading, Mass.: Addison Wesley, 1973), 95. Marx uses the phrase to refer to the disruption of a pastoral interlude, typically by a machine (e.g., the locomotive whose whistle pierces the quiet of Walden Pond).

31. Keith Thomas discusses the evolution of attitudes toward the creatures (in England) in *Man and the Natural World* (New York: Pantheon, 1983).

32. Leopold, *Sand County Almanac,* 36: "The woodcock is a living refutation of the theory that the utility of a game bird is to serve as a target, or to pose gracefully on a slice of toast. No one would rather hunt woodcock in October than I, but since learning of the sky dance I find myself calling one or two birds enough."

33. Macgillivray revised this passage to read: "It is a sorrowful sight after all: see that poor thing gasping hard in the agonies of death, its legs quivering with convulsive twitches, its bright eyes fading into glazed obscurity" (*OB,* 3:37). Audubon's manuscript can be found in the Beinecke Library collection.

34. For a more dispassionate account of rail shooting, in this case on the Delaware River near Philadelphia, see Audubon's predecessor Alexander

Wilson. T. M. Brewer, ed., *Wilson's American Ornithology* (1839; New York: H. S. Samuels, 1852). Wilson begins by observing that of all land and water fowl "perhaps none afford the sportsman more agreeable amusement, or a more delicious repast" (418).

35. Maria R. Audubon, *Audubon and his Journals*, 1:509; 2:131.

36. Alton A. Lindsey, ed., *Bicentennial of John James Audubon* (Bloomington: Indiana University Press, 1985), 119. The *Missouri River Journals* were written in 1843 but were not published until 1897, in Maria E. Audubon, *Audubon and His Journals* (New York: Charles Scribner's Sons, 1897).

37. Quoted in Lindsey, *Bicentennial of John James Audubon,* 121.

38. Angela Miller comments on the sense of literary plot in contemporary discussions of nineteenth-century landscape painting, specifically the movement from wilderness through energetic settlement to resolution in the creation of "a new mythic republic." See *The Empire of the Eye* (Ithaca: Cornell University Press, 1993), 83.

39. Annette Kolodny sees "a disturbed and disturbing ambivalence" in Audubon's evocation of the Ohio valley in its earlier state and various evidences of guilt in Audubon's descriptions of the exploitation of the land in his episodes. See *The Lay of the Land* (Chapel Hill: University of North Carolina Press, 1975), 74–89. Michael Branch comments on the elegiac character of "The Ohio" and the awareness Audubon shows of his role in documenting a vanishing American wilderness. See "Indexing American Possibilities," 293–94, 296.

40. Ford, *1826 Journal of John James Audubon,* 285. David Mazel suggests that Audubon saw Scott, and Cooper, as capable of cultural accomplishments that would offset the loss of wilderness. See *American Literary Environmentalism* (Athens: University of Georgia Press, 2000), 64–65.

Chapter 2

1. Thoreau, *Excursions,* 103. Thoreau's citations of Audubon demonstrate a familiarity with both the *Ornithological Biography* and *The Quadrupeds of North America,* the latter based on Audubon's journey up the Missouri River. He refers in his journal to a stop in Boston to see an exhibition of Audubon's *Birds of America.* Henry David Thoreau, *Journal: 1837–44* (Princeton: Princeton University Press, 1981–), 5:70.

2. Thoreau, *Excursions,* 125.

3. Thoreau, *Excursions,* 105, 114.

4. David R. Foster, *Thoreau's Country* (Cambridge: Harvard University Press, 1999).

5. Foster, *Thoreau's Country,* 222.

6. Thoreau, *Journal: 1837–44,* Bradford Torrey and Francis H. Allen,

eds., *The Journal of Henry D. Thoreau,* 2 vols. (Boston: Houghton Mifflin, 1906; New York: Dover, 1962), 2:985 [8:221].

7. Thoreau, *Journal: 1837–44,* 1:115.

8. Bradford Torrey and Francis H. Allen, eds., *The Journal of Henry David Thoreau* (Boston: Houghton Mifflin, 1906), 2:1063 [9:42–43].

9. I am thinking expecially of Frederick Garber, *Thoreau's Redemptive Imagination* (New York: New York University Press, 1977), and James McIntosh, *Thoreau as Romantic Naturalist* (Ithaca: Cornell University Press, 1974).

10. For a useful article that addresses this topic directly, see Jonathan Fairbanks, "Thoreau: Speaker for Wildness," *South Atlantic Quarterly* 70 (1971): 485–506.

11. Thoreau, *Journal: 1837–44,* 1:308.

12. Thoreau, *Journal: 1837–44,* 4:480, 4:320.

13. Thoreau, *Journal: 1837–44,* 4:310.

14. It functions in a similar way for Gary Snyder in his poem about the evolution of his home environment in the Sierra foothills, "What Happened Here Before," which concludes with the line "Blue jay screeches from a pine," juxtaposed with the image of military jets roaring overhead. Gary Snyder, *Turtle Island* (New York: New Directions, 1974), 81.

15. Thoreau, *Journal: 1837–44,* 3:73.

16. Thoreau, *Journal: 1837–44,* 2:64.

17. Thoreau, *Journal: 1837–44,* 4:460.

18. Thoreau, *Journal: 1837–44,* 4:307. Lawrence Buell discusses Thoreau's aesthetic sense of landscape and his manipulation of the picturesque. See *Environmental Imagination,* 131, 408–11. For an examination of the influence of Gilpin and Ruskin on Thoreau's "aesthetic of nature" see H. Daniel Peck, *Thoreau's Morning Work* (New Haven: Yale University Press, 1990), 49–50, 65, 81, 110.

19. Henry David Thoreau, *A Week on the Concord and Merrimack Rivers,* ed. Carl F. Hovde (1849; Princeton: Princeton University Press, 1980), 171.

20. Thoreau, *Journal: 1837–44,* 1:495.

21. Thoreau, *Journal: 1837–44,* 2:179. Thoreau's reworking of this passage in *Walden* sharpens the prose but omits elements that are interesting for my purposes: "Grow wild according to thy nature, like these sedges and brakes, which will never become English hay." Henry David Thoreau, *Walden,* ed. J. Lyndon Shanley (1854; Princeton: Princeton University Press, 1971), 207.

22. Thoreau, *Excursions,* 229.

23. Thoreau, *Journal: 1837–44,* 5:226. William Howarth discusses the significance of swamps for Thoreau, with particular reference to a series of visits to Beck Stow's swamp in the spring of 1858. See "Imagined Territory: The Writing of Wetlands," *New Literary History* 30 (1999): 826–88. See also

David C. Miller, *Dark Eden: The Swamp in Nineteenth-Century American Culture* (Cambridge: Cambridge University Press, 1989), 214–23.

24. Thoreau, *Journal: 1837–44*, 1:320, entry for 13 August 1841; Thoreau, *Excursions*, 125.

25. Thoreau, *Journal: 1837–44*, 5:281.

26. Thoreau, *Journal: 1837–44*, 4:23, 5:224.

27. Thoreau, *Journal: 1837–44*, 5:323.

28. Thoreau, *Journal: 1837–44*, 1:129.

29. Thoreau, *Journal: 1837–44*, 3:97.

30. Thoreau, *Walden*, 317.

31. Thoreau, *Journal: 1837–44*, 4:23–24.

32. Thoreau, *Journal: 1837–44*, 5:302.

33. Thoreau, *Journal: 1837–44*, 9:182; Torrey and Allen, *Journal of Henry D. Thoreau*, 2:1098.

34. I am indebted to Robert Sayre's comprehensive *Thoreau and the American Indians* (Princeton: Princeton University Press, 1977). See 213–14 for a summary of the evolution of Thoreau's views.

35. Thoreau, *A Week*, 4.

36. Thoreau, *A Week*, 69.

37. Thoreau, *A Week*, 67, 66.

38. Quoted in Sayre, *Thoreau and the American Indians*, 60.

39. Thoreau, *Journal: 1837–44*, 3:231. Thoreau proclaimed in an 1841 letter to Lucy Brown: "I grow savager and savager every day, as if fed on raw meat." Carl Bode and Walter Harding, eds., *The Correspondence of Henry David Thoreau* (New York: New York University Press, 1958), 45.

40. Thoreau, *A Week*, 456.

41. Thoreau, *Walden*, 210.

42. "I fear that we are such gods or demigods only as fauns or satyrs, the divine allied to beasts" (*Walden*, 220). See Elisa New's reading of "Higher Laws" in *The Line's Eye* (Cambridge: Harvard University Press, 1998), 111–15.

43. Thoreau, *Walden*, 210.

44. "Once I went so far as to slaughter a woodchuck which ravaged my bean-field . . . and devour him, partly for experiment's sake." Thoreau, *Walden*, 59.

45. *MW*, 206.

46. Thoreau, *Excursions*, 313.

47. Thoreau, *Excursions*, 311.

48. Thoreau, *Journal: 1837–44*, 4:407.

49. Thoreau, *Walden*, 234–36.

50. Thoreau, *Excursions*, 301.

51. Henry David Thoreau, *Cape Cod*, ed. Joseph J. Moldenhauer (1865; Princeton: Princeton University Press, 1988), 90, 148.

52. *MW*, 11.

53. Gerard Manley Hopkins, "Inversnaid," in *The Poetical Works of Gerard Manley Hopkins,* ed. Norman H. Mackenzie (Oxford: Clarendon Press, 1990), 168.

54. Thoreau, *A Week,* 107, 396.

55. The point is Garber's. See *Thoreau's Redemptive Imagination,* 90–91.

56. Joseph Moldenhauer offers a brief summary of the evolution of these views in *"The Maine Woods,"* in *The Cambridge Companion to Henry David Thoreau,* ed. Joel Myerson (Cambridge: Cambridge University Press, 1995), 139–40. Ronald Wesley Haug, "The Mark of the Wilderness in Thoreau's Contact with Ktaadn," *Texas Studies in Literature and Language* 24 (1982): 23–46, argues that Thoreau responds with religious awe and a Burkean sense of the sublime. Steven Fink, *Prophet in the Marketplace,* 177ff., sees Thoreau as drawing upon conventions of the romantic sublime and finding various ways of sustaining his verbal mastery. For a reading of the episode as a failure of the imagination, see Garber, *Thoreau's Redemptive Imagination,* 81ff. McIntosh, *Thoreau as Romantic Naturalist,* 198–210, sees the episode as representing a personal and metaphysical shock, followed by withdrawal. John Tallmadge stresses the resistance of Thoreau's subject matter and his failure to find adequate ways of responding to it. See "'Ktaadn': Thoreau in the Wilderness of Words," *ESQ* 31 (1985): 137–48.

57. See Stephen Adams and Donald Ross Jr., *Revising Mythologies: The Composition of Thoreau's Major Works* (Charlottesville: University of Virginia Press, 1988), chapter 5, and Randall Roorda, *Dramas of Solitude* (Albany: SUNY Press, 1998), chapter 2. Roorda demonstrates that Thoreau's account of his experience on Katahdin is a "narrative construct" rather than anything representing a transcription of field notes. He shows, among other things, how Thoreau produced a climax by moving the reflections associated with the Burnt Lands (*"Contact! Contact!"*) from another section of an early draft.

58. Thoreau, *A Week,* 245–46.

59. John Milton, *Paradise Lost,* in *John Milton: Complete Poems and Major Prose,* ed. Merritt Hughes (Indianapolis: Bobbs Merrill Co., 1957), 254.

60. Milton, *Paradise Lost,* 255.

61. See Nash, *Wilderness and the American Mind,* 102–3. Nash cites George Catlin's 1832 suggestion of a *"Nation's Park"* to preserve Indians, buffaloes, and wilderness environment (101).

62. Thoreau, *Journal: 1837–44,* 4:47, entry for 5 September 1851.

63. Thoreau, *Journal: 1837–44,* 3:251.

64. Thoreau, *Walden,* 317.

65. See Adams and Ross, *Revising Mythologies,* 145–46. Thoreau drew upon notes for two lectures, "The Wild" (1851) and "Walking" (1856), as well as upon journal entries.

66. Thoreau, *Excursions,* 226.

67. See especially David M. Robinson, "Thoreau's 'Walking' and the Ecological Imperative," in *Approaches to Teaching Thoreau's 'Walden' and Other Works,* ed. Richard J. Schneider (New York: Modern Language Association, 1996), 169–74; and William Rossi, "'The Limits of an Afternoon Walk': Coleridgean Polarity in Thoreau's 'Walking,'" *ESQ* 33 (1987): 94–109. Robinson describes the wild as "from the first defined by civilization and thereby bounded by the very system of values it attempts to impose" (172). Rossi argues that the interaction of the wild and the civilized is productive because they are polar opposites and identifies the fertility of Thoreau's wild with the "generative potential" it embodies "for an interacting consciousness" (98).

68. Bunyan's Christian must abandon his family to seek the New Jerusalem. The underlying biblical text is Luke 14:26: "If any man come to me, and hate not his father, and mother, and wife, and children, and brethren, and sisters, yea, and his own life also, he cannot be my disciple."

69. See Garber, *Thoreau's Redemptive Imagination,* 219–20; Rossi, "'Limits of an Afternoon Walk,'" 104–5; Robinson, "Thoreau's 'Walking' and the Ecological Imagination," 174.

70. Thoreau never revised (or published) "Huckleberries" after completing a draft in 1861. See William Howarth, *The Book of Concord* (New York: Viking, 1982), 200.

71. See Robert Sattelmeyer, ed., *The Natural History Essays* (Salt Lake City: Peregrine Smith Books, 1980), for a good introduction to these essays. John Hildebiddle, *Thoreau: A Naturalist's Liberty* (Cambridge: Harvard University Press, 1983), considers them in the context of Thoreau's interest in natural history.

72. See Fink, *Prophet in the Marketplace,* 276ff.

73. Thoreau, *Excursions,* 312.

74. Apparently based upon Sal Cummings, a country woman "conversant with nuts and berries." Torrey and Allen, *Journal of Henry D. Thoreau,* 2:1213 [10:124].

75. Thoreau, *Excursions,* 321.

76. Thoreau, *Excursions,* 322. Fink, *Prophet in the Marketplace,* 284, finds in Thoreau's appropriation of Joel a "witty mock seriousness" appropriate to the personal essay. I would prefer to think that Thoreau saw that he could have it both ways, recognizing the irony in his identification with Joel but using Joel's prophetic voice to give greater seriousness to his complaint.

77. Henry David Thoreau, *Huckleberries,* ed. Leo Stoller (Iowa and New York: The Windhover Press of the University of Iowa and The New York Public Library, 1970), 29.

78. Thoreau, *Huckleberries,* 25.

79. Thoreau, *Huckleberries,* 23.

80. Thoreau, *Huckleberries,* 28.

81. Thoreau, *Excursions,* 270. See Robert Milder, *Reimagining Thoreau* (Cambridge: Cambridge University Press, 1995), 186–90.

82. Thoreau, *Excursions,* 274.

83. Thoreau, *Excursions,* 280.

84. Thoreau, *Excursions,* 284.

85. Thoreau, *Excursions,* 280.

86. Thoreau, *Huckleberries,* 35. The passage, which I have truncated here, appears as the conclusion to Thoreau's first draft of a lecture on "Huckleberries," according to Stoller.

87. Henry David Thoreau, *Wild Fruits,* ed. Bradley F. Dean (New York: W. W. Norton, 2000). This unfinished manuscript served as the source of both "Wild Apples" and "Huckleberries."

88. Thoreau, *Wild Fruits,* 19.

89. Thoreau, *Wild Fruits,* 3.

90. Thoreau, *Wild Fruits,* 168. Dean comments on this passage in his introduction, xvi.

91. Thoreau, *Wild Fruits,* 217.

92. Thoreau, *Wild Fruits,* 106.

Chapter 3

1. See Catherine L. Albanese, *Nature Religion in America* (Chicago: University of Chicago Press, 1990), 93–105, for a discussion of Muir's relationship to the transcendentalism of Emerson and Thoreau.

2. I have examined those in the Holt-Atherton Department of the University of the Pacific Libraries. They include individual editions of *Walden* (1862), *Excursions* (1863), *The Maine Woods* (1868), *A Week on the Concord and Merrimack Rivers* (1868), and the complete *Walden* edition (1906). Muir also owned the Riverside edition of the complete works, now in the Huntington Library. Muir indexed and sometimes copied passages on the end pages of his volumes and marked passages in the text with vertical lines, occasionally adding marginal notes. Richard Fleck has analyzed Thoreau's influence on Muir in "John Muir's Homage to Henry David Thoreau," *The Pacific Historian* 29 (1985): 55–64.

3. In his copy of the 1906 edition of *The Maine Woods* Muir responded to many of the same manifestations of wildness and showed more interest in Thoreau's descriptions of the devastations of logging in "The Allegash and the East Branch."

4. William Frederic Badè, ed., *The Life and Letters of John Muir,* 2 vols. (Boston: Houghton Mifflin, 1923–24), 1:270–73.

5. "Heaven knows that John Baptist was not more eager to get all his

fellow sinners into the Jordan than I to baptize all of mine in the beauty of God's mountains." Linnie Marsh Wolfe, ed., *John of the Mountains: The Unpublished Journals of John Muir* (Madison: University of Wisconsin Press, 1979), 86.

6. Muir's original journal for the summer of 1869 has been lost. In 1887 he produced three notebooks (labeled "Sierra Journal of 1869," vols. 1, 2, and 3) that represent the experience of that summer. He revised these extensively in preparing the book for publication in 1911. The notebooks, with typed transcripts incorporating the revisions that he made, can be found in the John Muir Papers at the University of the Pacific. Steven J. Holmes has argued convincingly that the 1887 notebooks constitute a substantially revised and expanded version of the missing 1869 journal. He concludes, from a comparison of the notebooks with extant letters and journals from before and after the summer of 1869, that the record of places and objects presented in the notebooks is largely reliable but that "much of the narrative structure and interpretive language" reflects extensive revision between 1869 and 1887. See *The Young John Muir* (Madison: University of Wisconsin Press, 1999), app. A, 253–59.

7. John Muir, *The Mountains of California* (1894; New York: Viking, 1985), 40.

8. John Muir Papers, draft of "The Douglas Squirrel," ca. 1876.

9. Thoreau, *Excursions,* 226.

10. See the epigraph to this chapter, a 1912 fragment preserved in the John Muir Papers.

11. For an extended comparison of the two episodes see Edgar M. Castellini, "On the Tops of Mountains: John Muir and Henry Thoreau," in *John Muir: Life and Work,* ed. Sally M. Miller (Albuquerque: University of New Mexico Press, 1993), 152–66. Castellini stresses the similarities between the scenes, whereas I find their differences more illuminating.

12. A draft of what he then called "In the Heart of the California Alps" (John Muir Papers, ca. 1876) contains three different versions of one page (47) in which Muir tries out ways of representing the harmonies and the animation that he found in "such all-embracing alp-top visions."

13. Michael Cohen comments that Muir sees "not a map, but a manifestation of flow." Cohen offers an extended reading of "A Near View of the High Sierra," finding it a key text for his explanation of Muir's spiritual awakening through the sense of loss and recovery of self he experiences on the cliff face. See *The Pathless Way: John Muir and American Wilderness* (Madison: University of Wisconsin Press, 1984), 67–81.

14. John Muir, *My First Summer in the Sierra* (1911; New York: Viking, 1987), 236.

15. The Thoreau quotation is from *Excursions,* 232.

16. Sherman Paul comments on Muir's use of present participles to convey a sense of nature as process. See *For Love of the World,* 253.

17. Cohen, *Pathless Way,* 350–52.

18. From Muir's original version of the entry of 6 June ("How glorious a conversion"). "Sierra Journal," 1:28.

19. Albanese, *Nature Religion in America,* 96.

20. Wolfe, *John of the Mountains,* 118.

21. Muir, *Mountains of California,* 59.

22. Muir's responses to *The Maine Woods* (1868 edition) suggest that he found a version of this paradoxical doubleness of nature in Thoreau. He marks with double lines Thoreau's "Who shall describe the inexpressible tenderness and immortal life of the grim forest?" One of his endnotes reads, "stern yet gentle furious untameable."

23. Wolfe, *John of the Mountains,* 203.

24. John Muir, *The Yosemite* (1912; Madison: University of Wisconsin Press, 1987), 85–86.

25. Muir, *Mountains of California,* 173.

26. Wolfe, *John of the Mountains,* 317.

27. Michael Cohen faults Muir for a lack of understanding of population dynamics and the role of predation and for an avoidance of evidence of strife in nature. See *Pathless Way,* 166–68, 180.

28. A sentence that Muir omitted from the original version of this entry shows him using other favorite metaphors (dance, music, flood) to express his sense of the vitality and flow of nature: "The finest particles of matter however compacted, grouped in crystal, or swirling free in winds are yet dancing in one grand music & the deadest stone is drenched with life & all things flow in one cosmic flood." "Sierra Journal," 1:116.

29. Compare Thoreau's question upon observing the difficulty of finding an unblemished lily pad: "Is not disease the rule of existence?" Thoreau, *Journal: 1837–44,* 4:26.

30. Wolfe, *John of the Mountains,* 93.

31. Muir illustrates nature's standard of cleanliness in "Wild Wool," in commenting on the "clothing" of "her beautiful wildlings": "No matter what the circumstances of their lives may be, she never allows them to go dirty or ragged. The mole, living always in the dark and in the dirt, is yet as clean as the otter or the wave-washed seal; and our wild sheep, wading in snow, roaming through bushes, and leaping among jagged storm-beaten cliffs, wears a dress so exquisitely adapted to its mountain life that it is always found as unruffled and stainless as a bird." William Cronon, ed., *John Muir: Nature Writings* (New York: The Library of America, 1997), 599–600.

32. See Mary Douglas, *Purity and Danger* (London: Routledge, 1966), especially chapter 2, on defining and excluding uncleanness.

33. John Tallmadge, "John Muir and the Poetics of Natural Conversion,"

North Dakota Quarterly 59 (1991): 75–77. See also Cohen, *Pathless Way,* 106–11, for a discussion of Muir's use of the metaphor of the Book of Nature.

34. See Cohen, *Pathless Way,* 236–51, for a discussion of the essays Muir wrote for *Picturesque California* and of his ability to play to the contemporary vogue for the picturesque.

35. Muir, *Mountains of California,* 90.

36. John Tallmadge reads *My First Summer* as spiritual autobiography, relating it to the literature of conversion, particularly Paul and Augustine. See "John Muir and the Poetics," 62–79.

37. Quoted in Ronald Engberg and Donald Wesling, *John Muir: To Yosemite and Beyond* (Madison: University of Wisconsin Press, 1980), 162.

38. Richard Poirier, ed., *Ralph Waldo Emerson* (Oxford: Oxford University Press, 1990), 7.

39. Muir, *Yosemite,* 255.

40. "Rocks have a kind of life perhaps not so different from ours as we imagine . . . their natural beauty is only a veil covering spiritual beauty—a divine incarnation—instonation." Quoted in Engberg and Wesling, *John Muir: To Yosemite and Beyond,* 113.

41. Wolfe, *John of the Mountains,* 216.

42. The other animal Muir wrote about at length is the wild sheep, the subject of a chapter in *The Mountains of California* and of the essay "Wild Wool." Wild sheep appealed to Muir as "animal mountaineers," capable of startling feats he could only envy, and as foils to the lethargic domestic sheep he saw destroying Sierra meadows. In "Wild Wool" he makes the fineness of their wool the basis for arguing the superiority of nature to culture.

43. Muir, *Mountains of California,* 191.

44. Muir, *Mountains of California,* 161.

45. Muir, *Mountains of California,* 145.

46. Muir, *Mountains of California,* 184, 185.

47. Muir, *Mountains of California,* 176.

48. Wolfe, *John of the Mountains,* 312.

49. Muir, *Mountains of California,* 179.

50. Cronon, *John Muir: Nature Writings,* 615.

51. Cronon, *John Muir: Nature Writings,* 613.

52. Cronon, *John Muir: Nature Writings,* 507.

53. Wolfe, *John of the Mountains,* 167.

54. Muir reshaped his account of his adventure on the brink of Yosemite Falls for dramatic effect. In revising, he omitted entirely a brief account of a prior visit to the top of the falls and moved his description of a night of "nerve strain," which followed it, to the next entry. In the journal version Muir describes himself edging down the slope at the top of the falls "on my back," controlling his speed in order to slip down "at my ease." These details,

demonstrating his caution, disappear in the printed version. Muir also omits a phrase about his motive for taking bitter artemisia leaves in his mouth ("hoping that they might keep me conscious of the body") before venturing onto the three-inch ledge from which he could look directly down into the falling water. Other revisions heighten the intensity of the prose. The "stream" becomes a "torrent"; snowy streamers "chant." Muir ends by resolving to keep away from "such extravagant, nerve-straining places," rather than simply "the head of the falls" and "the edge of the cliffs" as in the journal. See "Sierra Journal," 2:60, 2:62, 2:63.

55. Wolfe, *John of the Mountains,* 58. Undated fragment.

56. Wolfe, *John of the Mountains,* 43. Also see Cohen, *Pathless Way,* 137.

57. Muir, *Mountains of California,* 89.

58. Holmes, *Young John Muir,* 248. Holmes cites William Cronon's "The Trouble with Wilderness" as support for his effort to establish a new sense of Muir as a model for living with a nature not necessarily understood as wilderness.

59. Muir, *Mountains of California,* 90.

60. John Tallmadge describes the book as proceeding by "a series of small revelations that add up to a changed view of self and world." See "John Muir and the Poetics," 68.

61. In revising this passage Muir added more present participles and new descriptive detail, while dropping some elements of the original ("light as dry thistle down," "through groves of silver fir every needle shining"). "Sierra Journal," 3:129.

62. The "hitched" of the last phrase, which is frequently quoted, illustrates the older Muir's instinct for stylistic economy. His initial version has "bound fast by a thousand invincible cords that cannot be broken to everything in the universe." "Sierra Journal," 3:130.

63. See Michael Cohen, *Pathless Way,* 359–60, for a discussion of Muir's experiences on Cathedral Peak. Muir's original account of his time at the summit is more realistic and less charged with emotion than the printed version:

> How often I had gazed at it from our camps & from hills & ridge tops on my many short excursions. Now I had reached it, and I may say I have been at church today for the first time since leaving Wisconsin. I wandered over the roof wondering at the regularity and symmetry of its parts—scrambled around & up its spine & at the highest point ate my luncheon. But so high and giddy is the sheer precipitous front I found I could not swallow the bread while looking down.

In this version cassiope is simply a "long-looked for shrub." "Sierra Journal," 3:124–25.

64. See Kenneth R. Olwig's discussion of nineteenth-century perceptions of Yosemite in "Reinventing Common Nature: Yosemite and Mount Rushmore—A Meandering Tale of a Double Nature," in *Uncommon Ground: Toward Reinventing Nature*, ed. William Cronon (New York: Norton, 1995), 379–408.

Chapter 4

1. I am thinking primarily of Richard Shelton's astute comment: "*Desert Solitaire* was written by an archromantic trying desperately not to be romantic." See "Creeping Up on *Desert Solitaire*," in *Resist Much, Obey Little*, ed. James Hepworth and Gregory McNamee (Tucson: Harbinger House, 1985), 71–87. Shelton comments on the tension between the "cynically realistic" and the "soaring romantic" Abbey (75). Ann Ronald characterizes Abbey as a "writer of romance" who "[reshapes] his own desert universe into a mythic place" and discusses the "mythic" and "sacral" dimensions of his work. See *The New West of Edward Abbey* (Reno: The University of Nevada Press, 1982), 65. Patricia Limerick relates Abbey to the romantic tradition in *Desert Passages* (Albuquerque: University of New Mexico Press, 1985), 160–61. James McClintock has commented on the romantic element in Abbey and Jack London and sees them both as attracted to mystery. See *Nature's Kindred Spirits* (Madison: University of Wisconsin Press, 1994), 73–74.

2. Critics have described this tension variously. See David Copland Morris's illuminating analysis of the tension between what Morris describes as ironic and celebratory voices in passages from *Desert Solitaire*. "Celebration and Irony: The Polyphonic Voice of Edward Abbey's *Desert Solitaire*," *Western American Literature* 28 (1993): 21–32. Scott Slovic finds a similar division in the narrative voice of *Desert Solitaire* but characterizes this as a tension between "the yearning simultaneously to control and to surrender (or belong) to something beyond the self." See Slovic, *Seeking Awareness in American Nature Writing*, 96; see also chapter 4.

3. Lawrence Buell has described very well some of the forms this self-consciousness takes in *Desert Solitaire*, showing how the persona Abbey adopts in that work "repeatedly catches himself mythifying" and recognizes "how hard it is to part with the romantic furniture we say we want to jettison." See *Environmental Imagination*, 72–73.

4. Claire Lawrence, "'Getting the Desert into a Book': Nature Writing and the Problem of Representation in a Postmodern World," in *Coyote in the Maze*, ed. Peter Quigley (Salt Lake City: University of Utah Press), 158.

5. *DS*, 273.

6. Edward Abbey, *Down the River* (New York: E. P. Dutton, 1982),120.

7. See SueEllen Campbell, "Magpie," in *Coyote in the Maze,* ed. Peter Quigley (Salt Lake City: University of Utah Press), 33–46.

8. See Tom Lynch, "Nativity, Domesticity, and Exile in Edward Abbey's 'One True Home,'" in *Coyote in the Maze,* ed. Peter Quigley (Salt Lake City: University of Utah Press), 88–105. Lynch contrasts Abbey's vision of the Colorado Plateau with those of Leslie Marmon Silko, Simon Ortiz, and Luci Tapohonso, arguing that for them nature and culture are interwoven and that what Abbey sees as "alien" and "exotic" they find "a domestic place, a home full of family and friends" (93).

9. The question is SueEllen Campbell's. See "Magpie," 44.

10. Edward Abbey, introduction to *Eco-Defense: A Field Guide to Monkey-wrenching,* ed. Dave Foreman (Tucson: Earth First! Books, 1985), 31. In a journal entry for 1981 Abbey states that his concern for wilderness is "not aesthetic but physical, sensual, empathic, spiritual, but above all moral." Edward Abbey, *Confessions of a Barbarian* (Boston: Little Brown, 1994), 299. The entry goes on to take up the cause of the voiceless, including the mountain lion and the falcon, a cause Abbey began to promote in his published writing.

11. "The *Bloomsbury Review* Interview," by Dave Solheim and Rob Levin, in *Resist Much, Obey Little,* ed. James Hepworth and Gregory McNamee (Tucson: Harbinger House, 1985), 89–104.

12. Edward Abbey, *Abbey's Road* (New York: E. P. Dutton, 1979), 78.

13. Edward Abbey, *Beyond the Wall* (New York: Holt Rinehart, 1984), 55.

14. Edward Abbey, *A Voice Crying in the Wilderness* (New York: St. Martin's Press, 1989), 83.

15. Quoted in W. L. Rusho, *Everett Ruess: Vagabond for Beauty* (Salt Lake City: Peregrine Smith, 1983), 218.

16. See Paul Bryant's helpful discussion of Abbey's uses of the terms "paradox" and "bedrock." "The Structure and Unity of *Desert Solitaire," Western American Literature* 28 (1993): 3–19.

17. Robert Hass, ed., *Rock and Hawk: A Selection of Shorter Poems by Robinson Jeffers* (New York: Random House, 1987), 67, 167. James McClintock shows how Abbey echoes Jeffers here and in the "Bedrock and Paradox" chapter of *Desert Solitaire* and discusses their similarities and dissimilarities. See *Nature's Kindred Spirits,* 75–79. Diane Wakoski discusses Abbey's affinities with Jeffers's "inhumanism." See "Joining the Visionary 'Inhumanists,'" in *Resist Much, Obey Little,* ed. James Hepworth and Gregory McNamee (Tucson: Harbinger House, 1985), 117–23.

18. Abbey defends his attention to "the surface of things" in the introduction to *Desert Solitaire,* against imagined objections to a neglect of the

"underlying reality" (xi). Jeffers sought in his own way to get beyond surfaces. In "Oh, Lovely Rock" (*Rock and Hawk*) he describes feeling the "intense
reality" of the rock face under which he is camped and his sense of "seeing
through the / flame-lit surface into the real and bodily / And living rock"
(199). In "Rock and Hawk" (*Rock and Hawk*) he finds a "dark peace" (167) in
the rock. In both instances Jeffers endows the rock with a living, reassuring
presence.

19. Abbey gives this answer a comic turn in *The Monkey Wrench Gang*
(Philadelphia: Lippincott, 1975), when he shows Hayduke attributing it to
Doc Sarvis as he lies in wait for Bonnie's captors and adding, "But . . . what
about the smell of fear, Dad?" (239).

20. In another essay collected in *Abbey's Road*, "Science with a Human
Face," Abbey reacts against the fascination with Eastern religions and contrasts scientists he admires for contributing to knowledge (Darwin and Ein-
stein among them) with "the shamans, gurus, seers, and mystics of the earth,
East and West," claiming that he is holding up "the ragged flag of reason"
(125, 127).

21. John Van Dyke, *The Desert* (1901; Salt Lake City: Peregrine Smith,
1980) 106–7. Richard Shelton explores some of the similarities between
Abbey and Van Dyke. See "Creeping Up on *Desert Solitaire*," 81–86.

22. See Les Standiford, "Desert Places: An Exchange with Edward
Abbey," *Western Humanities Review* 24 (1970): 395–98.

23. In a late interview Abbey put the point even more strongly: "Writing
is a form of piety or worship. I try to write prose psalms that praise the
divine beauty of the natural world." In the same interview he responded to a
question about religion by saying: "If I have a religion, it's pantheism, the
belief that everything is in some sense holy, or divine, or sacred. Everything—even human life." See interview by Kay Jimerson, in *This Is About
Vision: Interviews with Southwestern Writers,* ed. William Balassi, John F. Craw-
ford, and Annie Eysturoy (Albuquerque: University of New Mexico Press,
1990), 53–57.

24. Abbey uses the word "sublime" in *Desert Solitaire* to suggest qualities
of the desert that defy categorization: "The desert lies beneath and soars
beyond any possible human qualification. Therefore, sublime" (219).

25. Sylvie Mathé discusses *Desert Solitaire,* and also Van Dyke's *The Desert,*
as a sustained "celebration amoureuse." See "Désir du désert: Hommage au
Grand Désert américain," *Revue Française D'Études Américaines* 50 (1991):
923–36. Richard Shelton finds Abbey, like Van Dyke, speaking of the desert
"like a man in love," particularly in the account of his departure at the end of
Desert Solitaire. See "Creeping Up on *Desert Solitaire*," 83–87. Randall Roorda
offers a critique of Shelton's use of the metaphor of romantic love, here and
especially in his introduction to Van Dyke's *The Desert.* See *Dramas of Solitude:*

Narratives of Retreat in American Nature Writing (Albany: State University of New York Press, 1998), 84, 240n.

26. Mary Austin, "The Land," in *Stories from the Country of Lost Borders,* ed. Marjorie Pryse (New Brunswick: Rutgers University Press, 1987), 160.

27. Edward Abbey, *Good News* (New York: E. P. Dutton, 1980), 60.

28. Abbey gave Hayduke a fantasy of Phoenix covered by sand dunes and "free men and wild women on horses, free women and wild men" roaming the canyonlands. *The Monkey Wrench Gang,* 100–101.

29. Van Dyke wrote in *The Desert* of the power of "shifting sands" to bury and destroy (28) and asserted that "sooner or later Nature will surely come to her own again. Nothing human is of long duration" (62).

Chapter 5

1. Wendell Berry, *Recollected Essays* (San Francisco: North Point Press, 1981), 273.

2. Berry, *Recollected Essays,* 313.

3. Wendell Berry, *Another Turn of the Crank* (Washington, D.C.: Counterpoint, 1995), 41.

4. Berry, *Another Turn of the Crank,* 314.

5. Wendell Berry, *Home Economics* (San Francisco: North Point Press, 1987), 139.

6. In "Preserving Wildness" Berry characterizes the fertile topsoil as "a dark wilderness, ultimately unknowable, teeming with wildlife" (*HE,* 140).

7. Wendell Berry, *The Long-Legged House* (New York: Harcourt, Brace and World, 1969), 101.

8. Herman Nibbelink discusses Berry's relationship to Thoreau, with particular reference to the sequence of poems published as *Clearing* (New York: Harcourt Brace Jovanovich, 1977). See "Thoreau and Wendell Berry: Bachelor and Husband of Nature," in *Wendell Berry,* ed. Paul Merchant (Lewiston, Idaho: Confluence Press, 1991), 135–51. Ted Olson argues that Berry diverges from Thoreau in his work since the mid-1970s, particularly in his defense of an agricultural way of life. See " 'In Search of a More Human Nature': Wendell Berry's Revision of Thoreau's Experiment," in *Thoreau's Sense of Place,* ed. Richard J. Schneider (Iowa City: University of Iowa Press, 2000), 61–69.

9. Scott Slovic discusses *The Long-Legged House* as a "celebration of watchfulness." See Slovic, *Seeking Awareness in American Nature Writing,* chapter 5.

10. Wendell Berry, *A Timbered Choir: The Sabbath Poems, 1979–1997* (Washington, D.C.: Counterpoint, 1998), 40.

11. Alan Rudrum, ed., *Henry Vaughan: The Complete Poems* (New Haven:Yale University Press, 1981), 289–90.

12. See David E. Gamble, "Wendell Berry: The Mad Farmer and Wilderness," *Kentucky Review* 2 (1988): 40–52. Gamble discusses the lessons of the experience Berry describes in *The Unforeseen Wilderness.*

13. Wendell Berry, *The Unforeseen Wilderness,* rev. ed. (San Francisco: North Point Press, 1991).

14. See John Mark Faragher, *Daniel Boone* (New York: Holt, 1992), for an excellent reconstruction of the life and an analysis of the growth of the legend. Berry draws upon an episode, reported in John Filson's *The Adventures of Col. Daniel Boone* (Norwich, Conn.: John Trumbull, 1786), in which Boone remained in Kentucky by himself while his brother went back to Virginia for supplies. In his early poem "Boone" (*CP,* 10), Berry presents a more complex image of Boone in old age, removed to Missouri, reflecting on the elusiveness of the dreams that brought him over the mountains in the first place: "The search / withholds the joy from what is found . . . . / There are no arrivals" (*CP,* 10).

15. Robert Collins has written suggestively on the strain of skepticism about language, as opposed to the "song of the world," that runs through Berry's poetry. See "A More Mingled Music: Wendell Berry's Ambivalent View of Language," *Modern Poetry Studies* 11 (1982): 35–56.

16. Leonard M. Scigaj has linked Berry's attraction to silence with Merleau-Ponty's sense of silence "as the world of our immediate, unreflective, unthematized perception." See Scigaj, *Sustainable Poetry,* 75, 135–37.

17. John Lang discusses Berry's use of metaphors of song and silence in the poetry and his use of darkness (and silence) to express the numinous, invoking Rudolph Otto's *The Idea of the Holy.* See "'Close Mystery': Wendell Berry's Poetry of Incarnation," *Renascence* 35 (1982): 258–68. See also Scigaj, *Sustainable Poetry,* 155, 168, on the significance of darkness in Berry's poetry.

18. Jack Hicks comments on the concluding scene of the novel in "Wendell Berry's Husband to the World: *A Place on Earth,*" in *Wendell Berry,* ed. Paul Merchant (Lewiston, Idaho: Confluence Press, 1991), 118–34. See also Gamble, "Mad Farmer and Wilderness," 50–51.

19. Wendell Berry, *A Place on Earth,* rev. ed. (San Francisco: North Point Press, 1983), 317.

20. Wendell Berry, *The Memory of Old Jack* (New York: Harcourt, 1974), 192.

21. Wendell Berry, *The Wild Birds* (San Francisco: North Point Press, 1985), 82.

22. Wendell Berry, *Remembering* (San Francisco: North Point Press, 1988), 122.

23. Berry, *Wild Birds,* 104.

24. Berry, *Clearing.*

25. Wendell Berry, *The Life Story of Jayber Crow, Barber, of the Port William Membership, as Written by Himself* (Washington, D.C.: Counterpoint, 2000).

26. Berry, *Timbered Choir,* 6.

27. Wendell Berry, *Sabbaths* (San Francisco: North Point Press, 1987).

28. "The Country of Marriage," *CP,* 146–47.

29. Berry, *Place on Earth,* 317; Berry, *Timbered Choir,* 43.

Chapter 6

1. Mary Oliver, "Among Wind and Time," *Sierra,* November/December 1991, 34.

2. Mary Oliver, *Dream Work* (New York: The Atlantic Monthly Press, 1986).

3. See Robin Riley Fast's discussion of this and other poems dealing with Native Americans or reflecting Native influences. "The Native American Presence in Mary Oliver's Poetry," *The Kentucky Review* 13 (1993): 59–68.

4. Oliver evokes the spirit of Western adventuring in "Poem for My Father's Ghost," also from *Twelve Moons* (Boston: Little Brown, 1979), 35, in which she imagines her father as freed to be a "traveler" by death. He becomes one of the "bold men" he admired, striding through the Dakotas into the mountains to be welcomed to the last campfire in the "final hills" by "chieftains, warriors, and heroes."

5. *MW,* 78.

6. In a related poem, "The Fence," Oliver questions the way a fence creates a "myth of difference" between wilderness and "civilized" territory. *The River Styx, Ohio* (New York: Harcourt, Brace Jovanovich, 1972), 30.

7. Vicki Graham discusses the process by which Oliver identifies with bears in this sequence, invoking the concept of "sympathetic magic." See "'Into the Body of Another': Mary Oliver and the Poetics of Becoming Other," *Papers in Language and Literature* 30 (1994): 100.

8. Oliver anticipates "Blossom" in "Pink Moon—The Pond" (*Twelve Moons,* 7–8), in which she describes herself walking into the spring pond and identifying with the life of the frogs.

9. Mary Oliver, *Winter Hours* (Boston: Houghton Mifflin, 1999), 100.

10. Janet McNew supports her claim that for Oliver "physicality . . . becomes the most visionary spirituality" primarily by appealing to poems about eating and about sexual appetite in *American Primitive.* See "Mary Oliver and the Tradition of Romantic Nature Poetry," *Contemporary Literature* 30 (1989): 59–77.

11. Interview by Mindy Weinstein, in *Our Other Voices,* ed. John Wheatcroft (Lewisburg, Penn.: Bucknell University Press, 1991), 140–51.

12. Oliver describes deadly mushrooms as "full of paralysis" ("Mushrooms," *AP,* 4–5). In "Blossom" she recognizes "that time / chops at us all like an iron / hoe, that death / is a state of paralysis."

13. Oliver comments on the restless enthusiasm of Muir's *Travels in Alaska.* See "Four Companions with a Zest for Life," in *Blue Pastures* (New York: Harcourt Brace, 1995), 34–35.

14. "Maples," *WW,* 20.

15. "I would talk about the owl and the thunderworm and the daffodil and the red-spotted newt as a company of spirits, as well as bodies." Oliver, *Winter Hours,* 102.

16. Oliver, *Winter Hours,* 99.

17. Cf. Whitman in "To Think of Time": "I swear I think now that every thing without exception has an eternal soul! / The trees have, rooted in the ground! the weeds of the sea have! the animals!" *Leaves of Grass,* ed. Harold W. Blodgett and Sculley Bradley (New York: New York University Press, 1965), 440.

18. "The Kingfisher," *HL,* 18; "This Morning Again It Was in the Dusty Pines," *NSP,* 24.

19. I am indebted to Scott Bryson's paper on Oliver at the June 1999 conference of the Association for the Study of Literature and the Environment.

20. "Staying Alive," *BP,* 64. Oliver describes this state as one she finds "deep inside books" as well as "out in the fields."

21. Oliver returns to this moment in *Winter Hours* (96): "Here is where two deer approached me one morning, in an unforgettable sweetness, their faces like light brown flowers, their eyes kindred and full of curiosity. The mouth of one of them, and its vibrant tongue, touched my hand."

22. Mary Oliver, *White Pine* (New York: Harcourt, Brace, & Co., 1984).

23. Cf. Thoreau's comments on "the art of spending a day" in his journal entry for 7 September 1851, part of a meditation on "how to extract its honey from the flower of the world": "it behooves us to be attentive. If by watching all day & all night—I may detect some trace of the Ineffable—then will it not be worth the while to watch?" Thoreau, *Journal: 1837–44,* 4:53.

24. Oliver, *Winter Hours,* 73, 100.

25. "My Friend Walt Whitman" and "The Poet's Voice," *BP,* 13–16, 95–115; "Some Thoughts on Whitman," *Winter Hours,* 62–73.

26. *BP,* 97, 102. Oliver dedicated *American Primitive* to the memory of Wright, and his influence on her voice deserves more attention than I can give it here. For examples of Wright poems that suggest Oliver's stylistic habits, see *To a Blossoming Pear Tree* (New York: Farrar, Straus, and Giroux, 1977), in particular "Redwings," "The Best Days," "To a Blossoming Pear

Tree," and "Beautiful Ohio." I am indebted to my colleague Linda Gregerson for directing me to this book.

27. Mary Oliver, *The Leaf and the Cloud* (Cambridge, Mass.: Da Capo Press, 2000), 14.

28. *Walt Whitman: Complete Poetry and Collected Prose* (New York: Library of America, 1982), 27. Quoted in Oliver, *Winter Hours*, 65.

29. Oliver, *Winter Hours,* 107, 101.

30. Oliver, *Winter Hours,* 108.

31. *LC,* 35.

32. "Black Oaks," *WW,* 5.

33. Oliver, *Winter Hours.*

Conclusion

1. Berry, *Long-Legged House,* 200.
2. *MW,* 27–28.
3. Berry, *Timbered Choir,* 14.
4. "Work," *LC,* 15.

Index